Ergebnisse der Mathematik und ihrer Grenzgebiete

Band 79

Herausgegeben von P. R. Halmos · P. J. Hilton
R. Remmert · B. Szőkefalvi-Nagy

Unter Mitwirkung von L. V. Ahlfors · R. Baer
F. L. Bauer · A. Dold · J. L. Doob · S. Eilenberg
K. W. Gruenberg · M. Kneser · G. H. Müller
M. M. Postnikov · B. Segre · E. Sperner

Geschäftsführender Herausgeber: P. J. Hilton

A.V. Skorohod

Integration in Hilbert Space

Translated from the Russian by
Kenneth Wickwire

Springer-Verlag
Berlin Heidelberg New York 1974

A. V. Skorohod
Institute of Mathematics AN USSR, Kiev

Title of the Russian Edition:

Integrirovanie v gilbertovyh prostranstvah
Publisher: Nauka Izdatel'stvo 1974

Translator: Kenneth Wickwire

Statistical Laboratory, Department of Mathematics, University of Manchester, Manchester 13, U K

AMS Subject Classification (1970):
28-02, 28A10, 28A25, 28A30, 28A35, 28A40, 28A65, 46C10, 60B05, 60B10, 60E05

ISBN-13: 978-3-642-65634-7 e-ISBN-13: 978-3-642-65632-3
DOI: 10.1007/978-3-642-65632-3

Softcover reprint of the hardcover 1st edition 1974

A.V. Skorohod

Integration in Hilbert Space

Translated from the Russian by
Kenneth Wickwire

Springer-Verlag
New York Heidelberg Berlin 1974

A. V. Skorohod
Institute of Mathematics AN USSR, Kiev

Title of the Russian Edition:

Integrirovanie v gilbertovyh prostranstvah
Publisher: Nauka Izdatel'stvo 1974

Translator: Kenneth Wickwire
Statistical Laboratory, Department of Mathematics,
University of Manchester, Manchester 13, U K

AMS Subject Classification (1970):
28-02, 28A10, 28A25, 28A30, 28A35, 28A40, 28A65
46C10, 60B05, 60B10, 60E05

ISBN-13: 978-3-642-65634-7 e-ISBN-13: 978-3-642-65632-3
DOI: 10.1007/978-3-642-65632-3

Softcover reprint of the hardcover 1st edition 1974

Author's Preface

Integration in function spaces arose in probability theory when a general theory of random processes was constructed. Here credit is certainly due to N. Wiener, who constructed a measure in function space, integrals with respect to which express the mean value of functionals of Brownian motion trajectories. Brownian trajectories had previously been considered as merely physical (rather than mathematical) phenomena. A. N. Kolmogorov generalized Wiener's construction to allow one to establish the existence of a measure corresponding to an arbitrary random process. These investigations were the beginning of the development of the theory of stochastic processes. A considerable part of this theory involves the solution of problems in the theory of measures on function spaces in the specific language of stochastic processes. For example, finding the properties of sample functions is connected with the problem of the existence of a measure on some space; certain problems in statistics reduce to the calculation of the density of one measure w. r. t. another one, and the study of transformations of random processes leads to the study of transformations of function spaces with measure. One must note that the language of probability theory tends to obscure the results obtained in these areas for mathematicians working in other fields.

Another direction leading to the study of integrals in function space is the theory and application of differential equations. A. N. Kolmogorov has shown a connection between the solutions of second-order parabolic differential equations and the means of certain stochastic processes (i. e., integrals w. r. t. measures in function space). Essential progress was made here by M. Kac, who expressed the solution of the equation

$$u_t = u_{xx} + v\,u$$

by means of an integral w. r. t. Wiener measure.

However, the most significant step in this direction was taken by R. Feynman who put his "continual integral" (the Feynman integral) at the basis of the structure of quantum mechanics. In particular, with the help of such an integral he was able to express the solution of the Schrödinger equation. The Feynman integral differs from the Wiener integral in that there exists no completely-additive measure with respect to which it can be written, so that the problem of the existence of the Feynman integral is a difficult one and has not yet been solved. The mathematical aspects of the problem are treated in the survey by Gel'fand and Yaglom [1]. The continual integral was subsequently used to investigate evolution equations of order greater than two. Unfortunately, it turned out that no measures in function space correspond to such equations, but rather quasi-measures, i. e., signed, finitely-additive set functions of unbounded variation. The deepest results here have been obtained by Ju. L. Daleckii (see, for example, his survey [1]).

In this book we consider only measures on function spaces which are taken to be separable Hilbert space. This is due to the fact that although Hilbert space is not essential in many cases, a number of important problems have been solved only for Hilbert space. To the latter is related the problem of the existence of a measure. At the same time, constructions carried out in Hilbert space can often be easily generalized to rather general linear spaces. There are remarks on this in the notes collected at the end of the book. Some justification for restricting ourselves to Hilbert space is also provided by the fact that Hilbert space has so far been sufficient for applications.

The author has set himself the goal of an orderly presentation in measure-theoretic language of the basic ideas of the theory of measure and integration in Hilbert Space, including those which have heretofore been available only in the theory of stochastic processes,. To the most important questions we consider are related: 1) methods of defining a measure and conditions for its existence, 2) measurable functions on Hilbert space with measure, 3) the construction of systems of orthogonal functions, 4) the absolute continuity of measures and the calculation of the density of one measure with respect to another, 5) the theory of quasi-invariant measures, 6) transformations of measures under transformations of the space and 7) surface integrals and Green's formula in Hilbert space. A considerable part of this material is published here for the first time.

In bibliographic notes collected at the end of the book we have attempted to clarify the role played by various authors in the development of the above ideas.

A. V. Skorohod

Translator's Preface

One of the satisfying aspects of mathematics is the ease (relative to other human pursuits) with which progress in certain of its tangled branches can be summarized and unified. That this is still not *Jedermanns Sache* is shown by the fact the subject of this book has waited fifty years for review and summary. To be sure, the introduction of Wiener's Differential Space in the early 20's was followed by a period in which it was viewed by many as a curiosity and neither understood nor appreciated. But the rapid development since 1923 of quantum physics and of the broad concept of "adaptive control" (two of the most exciting of the many directions taken by applied mathematics) and their increased (and occasionally incorrect) use of integration in function spaces has indicated that it is high time for such a book. The author is an expert on the subject whose various researches in the theory of stochastic processes have been closely connected with several areas of applied probability which are growing so rapidly that they often leave the rigorous justification of their techniques behind.

Skorohod has provided integration with respect to measures in function (Hilbert) space with a rigorous foundation in this important book. Through the good offices of Springer-Verlag and Nauka in Moscow I was able to begin translating it before its publication in the Soviet Union, which has advanced its availability in translation by several months.

I have corrected a number of misprints and typographical omissions and occasionally added what I believe to be a clarifying footnote. As usual, it was necessary to compile an index. Otherwise, this edition is the same as the original.

Thanks are due to Professor Skorohod for carefully reading the manuscript and thus clarifying many obscurities, and to my erstwhile companion M. for various favors.

Manchester, England, June 1974 K. Wickwire

Contents

Introduction

The general theory of measures has been constructed for arbitrary measurable spaces, i.e., for sets on which a σ-algebra of measurable subsets has been selected. It can thus be shown that when the measurable space is linear and the σ-algebra is related in a certain way to the algebraic structure of the space, then no special theory is necessary. This occurs if we restrict ourselves to a finite-dimensional space with a σ-algebra of Borel sets. Of course, in this case as well there arise special problems connected, for example, with invariant measures. However, no special theory is required for their solution. The situation changes considerably if we go over to infinite-dimensional spaces. For many important problems it is still impossible to give solutions as simple as those in the finite-dimensional case. Here are two examples of such problems. The first of these is that of defining a measure. In the finite-dimensional case, it is sufficient to define a measure on all parallelepipeds with sides parallel to the coordinate axes; the values of the measure on these sets are defined by some (distribution) function, so that to each such function there corresponds a measure. In the infinite-dimensional case this does not hold: there need not correspond a measure to each distribution function. The existence problem for a measure is far from being solved for all spaces. The second problem concerns conditions for the absolute continuity of measures and the form of the corresponding density. In the finite-dimensional case it is solved by differentiating the distribution function. In the infinite-dimensional case, the solution of this problem, even for concrete measures, is not at all trivial.

Hilbert space is the simplest and most natural generalization of a finite-dimensional space; in it are manifested all of the difficulties connected with an infinite number of dimensions. At the same time, the theory for this space has been most completely developed, so that a coherent presentation of it is already possible. The present book is devoted to this theory.

We list briefly the basic problems which will be treated and also mention the principal results.

Chapter 1 is devoted to methods of defining measures. The notions of finite-dimensional distribution, weak distribution and characteristic functional are defined and the Minlos-Sazonov theorem is proved, which gives necessary and sufficient conditions for the existence of a measure with given weak distribution (or characteristic functional). In this chapter we also define the class of Gaussian measures, which is important — especially for the theory of probability.

In the second chapter we consider measurable functions, in particular, linear and polynomial functions. It is necessary to note that a peculiarity of infinite-dimensional spaces is also manifested here: in the finite-dimensional case a measurable polynomial is necessarily continuous, but in Hilbert space this is not so. With the aid of a certain procedure which we describe it is possible to reduce the investigation of polynomial functions to the study of linear ones on some other space. We also consider various systems of functions which are orthogonal w.r.t. a given measure.

In Chapter 3 we study general questions of absolute continuity of measures in Hilbert space. As a preliminary we define and prove the existence of conditional measures and we also prove theorems on the convergence of martingales and semi-martingales. The general assumptions are adapted to product measures, Gaussian measures and mixed measures, i.e., measures obtained by mixing measures depending on a parameter and integrated w.r.t. such a parameter. We remark that a large number of papers of a probability-theoretic or applied character have been devoted to the question of the absolute continuity and singularity of Gaussian measures.

Chapter 4 investigates admissible shifts (translations) of a measure, i.e., shifts transporting a measure into another one which is absolutely continuous w.r.t. the original. The structure of the set of admissible shifts is studied and a condition for the admissibility of a shift is found in terms of the derivative of a measure w.r.t. a given direction. A peculiarity of infinite-dimensional space is the lack of a Lebesgue measure (invariant w.r.t. a shift) and even of a measure for which all shifts are admissible. Thus, measures are of interest for which there exists a sufficiently rich set of admissible shifts, for example a linear set, dense in the whole space. Such measures are said to be quasi-invariant. We will give a complete description of these.

Finally, in Chapter 5 we generalize some formulas of classical analysis to the infinite-dimensional case. The first of these is the substitution formula for integrals. In the case where the integral is taken w.r.t. a Gaussian measure, this topic has been the subject of lively discussion in the literature of probability theory for more than twenty years. Another problem relates to the construction of a surface integral connected with a measure which is not concentrated on the surface (exactly such a situation holds in the evaluation of the Lebesgue area of a surface from the Lebesgue volume it encloses.) We obtain a generalized Gauss formula for the construction of such a surface integral

The reader is expected to be acquainted with the basic theory of Hilbert space as well as that of measure and integration. In view of the large number of texts on these topics it will not be necessary to single out any for special recommendation; the author has endeavored to reduce to a minimum the use of definitions not given explicitly in the book as well as the number of unproved theorems. The notation is more or less customary making it unnecessary to offer any special explanations. References to the literature are not as a rule given in the text and are collected in notes at the end of the book.

Chapter I. Definition of a Measure in Hilbert Space

§ 1. Measurable Hilbert Spaces

Let X be a real separable Hilbert space with elements x, y, z, etc. Real numbers will be designated by small Greek letters; $\alpha x + \beta y$ and (x, y) will denote, as usual, the operations of multiplication of a vector (element of X) by a scalar, vector addition and the scalar product of vectors. The norm of a vector will be designated by

$$|x| = \sqrt{(x, x)} .$$

Subsets of X will be denoted by large Latin letters, classes of subsets by large Gothic letters. A class of sets $\mathfrak{A}$ in which we allow the operations of set difference, union and intersection is called a ring. A ring of sets $\mathfrak{A}$ containing X as an element is called an algebra. An algebra of sets in which the union operation can be applied countably many times is called a σ-algebra.

In the sequel the letter $\mathfrak{B}$ will denote the σ-algebra of Borel sets in X, i.e., the minimal σ-algebra of subsets of X containing all open sets (because of the separability of the space it is sufficient that the σ-algebra contain all spheres.) The measurable space $(X, \mathfrak{B})$, i.e., the set X with σ-algebra of measurable sets will be called a measurable Hilbert space.

We will study finite measures μ defined on $(X, \mathfrak{B})$. In this connection it is often convenient to assume that the measure μ is normalized, i.e., $\mu(X) = 1$. The primary aim of this chapter is to propose a method of constructing measures on $(X, \mathfrak{B})$. This method is essentially equivalent to the method of Lebesgue for constructing the measure on the line: first the measure is constructed on a certain class of elementary sets and then extended to the minimal σ-algebra containing these sets and completed. The latter operation (completion) will not be carried out since we only need the measure on $\mathfrak{B}$.

We will investigate certain classes of "simple" sets on which the values of the measure will be assigned. Let L be a finite-dimensional subspace of X, P_L the orthogonal projection operator on L and A a Borel set from L. A set of the form

$$\{x: P_L x \in A\}$$

will be called a cylinder set and the set A is called its base (we also say that this is a cylinder set with base in L). The class of all cylinder sets with bases in L is obviously a σ-algebra which we will write as $\mathfrak{B}^L$, whereby $\mathfrak{B}^L \subset \mathfrak{B}$. The union of all σ-algebras $\mathfrak{B}^L$ is an algebra. Indeed, if A_1 and A_2 belong to $\mathfrak{B}^{L_1}$ and $\mathfrak{B}^{L_2}$, then choosing $L = L_1 + L_2$ (the sum of the subspaces), we get

$$A_i \in \mathfrak{B}^L, i = 1, 2; \quad A_1 \cup A_2 \in \mathfrak{B}^L; \quad A_1 \cap A_2 \in \mathfrak{B}^L; \quad A_1 - A_2 \in \mathfrak{B}^L .$$

We denote this algebra of sets by $\mathfrak{B}_0$. It is called the algebra of cylinder sets. Sets from $\mathfrak{B}_0$ are also considered as "elementary" for the construction of a measure on $(X, \mathfrak{B})$. To convince ourselves that the value of the measure on $\mathfrak{B}_0$ uniquely defines that on $\mathfrak{B}$ it is necessary to show that the σ-closure of $\mathfrak{B}_0$ contains $\mathfrak{B}$ (i.e., $\mathfrak{B}$ is the smallest σ-algebra containing $\mathfrak{B}_0$). We will show this below and immediately note that the algebra $\mathfrak{B}_0$ still contains too many sets (X has too many finite-dimensional subspaces). It turns out that to define the measure it is sufficient to have a certain chain of increasing subspaces $L_n \subset L_{n+1}$ for which $\bigcup_n L_n$ is dense in X. Then $\bigcup_n \mathfrak{B}^{L_n} = \mathfrak{B}_0'$ will also be an algebra of sets. Since $\mathfrak{B}_0' \subset \mathfrak{B}_0$, it follows from the fact that the σ-closure of $\mathfrak{B}_0'$ coincides with $\mathfrak{B}$, that the σ-closure of $\mathfrak{B}_0$ coincides with $\mathfrak{B}$. Let us prove the first assertion. It is sufficient to show that the σ-closure of $\mathfrak{B}_0'$ contains any closed sphere S in X since an open sphere can be represented as a countable sum of increasing closed spheres. Set

$$S = \{x: |x - a| \leq \varrho\} .$$

Denote by P_n the operator projecting onto L_n and put

$$S_n = \{x: |P_n x - P_n a| \leq \varrho\} .$$

The set S_n belongs to $\mathfrak{B}^{L_n}$ and $S_n \supset S$. We will prove that

$$S = \bigcap_n S_n . \tag{1}$$

In fact, if $y \notin S$, then $|y - a| = \varrho + \delta , \delta > 0$.
But

$$\lim_{n\to\infty} P_n(y - a) = y - a$$

(in the sense of convergence in X). Thus, $|P_n(y - a)| \to |y - a|$ which means that for large enough n $|P_n(y - a)| = |P_n y - P_n a| > \varrho$, $y \notin S_n$. We have proved (1).

The advantage of the algebra $\mathfrak{B}_0'$ lies in the fact that it is the union of countably many σ-algebras of the form $\mathfrak{B}^L$. We now note that sets of the algebra $\mathfrak{B}^L$ are completely determined by Borel sets of the finite-dimensional Euclidean space L (L will be such a space if it is considered by itself and not as a subset of X). If L is an n-dimensional space, $\{e_k; k = 1, \ldots, n\}$ is an orthonormalized basis in L and $x = \sum \xi_k e_k$ is a decomposition of an arbitrary $x \in L$ w.r.t. this basis, then finite sums of sets of the form

$$\{x : \alpha_k \leq \xi_k \leq \beta_k;\ k = 1, \ldots, n;\ -\infty \leq \alpha_k < \beta_k \leq \infty\}$$

(these sets are called *rectangles*) generate an algebra of subsets of L, whose σ-closure coincides with the σ-algebra $\mathfrak{B}_L$ of Borel sets of L. The same property will be possessed by the algebra $\mathfrak{A}_L$, generated by rectangles with rational α_k and β_k. Let $\mathfrak{A}^L$ be the algebra of cylinder sets with bases in $\mathfrak{A}_L$. Then the σ-closure of $\mathfrak{A}^L$ coincides with $\mathfrak{B}^L$. Consequently, for an increasing sequence of subspaces L_n for which $\bigcup_n L_n$ is dense in X, the algebra

$$\mathfrak{A}_0' = \bigcup_n \mathfrak{A}^{L_n}$$

(we assume that bases in the L_n are chosen in a compatible way, i.e., the basis in L_{n+1} is obtained from that in L_n by adding the basis from the orthogonal complement to L_n in L_{n+1}; in this case $\mathfrak{A}_0'$ will actually be a set algebra since $\mathfrak{A}^{L_n} \subset \mathfrak{A}^{L_{n+1}}$) is such that its σ-closure contains $\mathfrak{B}$. Each of the algebras $\mathfrak{A}^{L_n}$ contains only a countable number of sets, which implies that $\mathfrak{A}_0'$ also contains a countable number of sets. We recall that the σ-algebra obtained from the σ-closure of a denumerable algebra of sets is said to be separable. By the same token, we have established that the σ-algebra $\mathfrak{B}$ is separable.

The algebras $\mathfrak{B}_0'$ and $\mathfrak{A}_0'$ make the measure problem easier since they contain fewer elementary sets. However, they depend on a certain set of finite-dimensional subspaces (and even on the bases in these subspaces), so that the definition of measures by means of these algebras possesses a non-invariant character. Hence, in those cases where it is necessary to describe the invariant properties of measures we will use the algebra $\mathfrak{B}_0$.

§ 2. Weak Distributions

Let μ be some normalized measure on $(X, \mathfrak{B})$. For each finite-dimensional subspace L of the space X we can consider the restriction of this measure to $\mathfrak{B}^L$. Now define the measure μ_L on the σ-algebra $\mathfrak{B}_L$ of Borel sets of L as follows:

$$\mu_L(A) = \mu(\{x : P_L x \in A\}),$$

where P_L is projection onto the subspace L. Then the fact that μ_L is a measure follows from the fact that for a sequence of non-overlapping sets A_n of $\mathfrak{B}_L$, the sets

$$P_L^{-1}\left(\bigcup_n A_n\right) = \bigcup_n P_L^{-1}(A_n),\ P_L^{-1}(A_n)$$

are also non-overlapping (here $P_L^{-1}(A)$ is the inverse image of the set A under the projection operator P_L, i.e., $P_L^{-1}(A) = \{x: P_L x \in A\}$). The measure μ_L is called the projection of the measure μ onto the subspace L.

Hence, with each measure μ we can associate the set of its projections $\{\mu_L\}$ on finite-dimensional subspaces of X. Obviously, knowing μ_L, one can determine μ on $\mathfrak{B}^L$. Hence, knowing μ_{L_n} for a sequence L_n of linear subspaces for which $L_n \subset L_{n+1}$ and $\bigcup_n L_n$ is dense in X, we can define μ on $\bigcup_n \mathfrak{B}^{L_n}$ and since the σ-closure of this algebra coincides with $\mathfrak{B}$ we can define μ on $\mathfrak{B}$ by the same token. This means that knowing $\{\mu_L\}$ or $\{\mu_{L_n}\}$, where L_n is the indicated sequence of subspaces, we can uniquely retrieve the measure.

The totality of all projections of a given measure are called the *finite-dimensional distributions* of the measure. The measures μ_{L_n} being projections of the same measure are compatible in a certain sense for different n. This compatibility condition follows from the fact that the base of a cylinder set is chosen non-uniquely. Let $L_1 \subset L_2$ and $A \in \mathfrak{B}_{L_1}$. Then the set $P_{L_1}^{-1}(A)$ can also be written as $P_{L_2}^{-1}(A_2)$, where $A_2 \in \mathfrak{B}_{L_2}$ is defined by the equality

$$A_2 = \{x: x \in L_2, P_{L_1} x \in A\} .$$

Since $P_{L_1}^{-1}(A) = P_{L_2}^{-1}(A_2)$, we have

$$\mu_{L_1}(A) = \mu(P_{L_1}^{-1}(A)) = \mu(P_{L_2}^{-1}(A_2)) = \mu_{L_2}(A_2)$$

Since $A_2 = P_{L_1}^{-1}(A) \cap L_2$, the compatibility condition can be written in the following form: for all $L_1 \subset L_2$ and $A \in \mathfrak{B}_{L_1}$

$$\mu_{L_1}(A) = \mu_{L_2}(P_{L_1}^{-1}(A) \cap L_2) . \tag{1}$$

The family of measures $\{\mu_L\}$, defined for all finite-dimensional subspaces L and satisfying the compatibility condition (1) is called a *weak distribution.*

If L_n is a sequence of subspaces, $L_n \subset L_{n+1}$, $\bigcup L_n$ dense in X and μ_{L_n} is a sequence of measures on $\mathfrak{B}_{L_n}$ satisfying the compatibility condition

$$\mu_{L_n}(A) = \mu_{L_{n+1}}(P_{L_n}^{-1}(A) \cap L_{n+1}) ,$$

then the sequence $\{\mu_{L_n}\}$ is called a *sequence of finite-dimensional distributions.*

From the results above it follows that to each measure on $(X, \mathfrak{B})$ there corresponds some weak distribution, and different weak distributions correspond to different measures. The problem of defining a measure with the help of weak distributions could now be solved quite easily if to each weak distribution therecorresponded some measure on $(X, \mathfrak{B})$. Unfortunately this is not the case. We will now derive conditions which must be satisfied by a weak distribution in order that some measure correspond to it.

Lemma 1. *Let S_ϱ be a sphere of radius ϱ: $S_\varrho = \{x: |x| \leq \varrho\}$. The weak distribution $\{\mu_L\}$ will be generated by some measure μ on $(X, \mathfrak{B})$ iff, for every $\varepsilon > 0$ there exists an $\eta > 0$ such that for all L*

$$\mu_L(S_\varrho \cap L) \geq 1 - \varepsilon \quad \text{when} \quad \varrho > \eta .$$

Proof. Necessity. If $\{\mu_L\}$ is generated by the measure μ, then choosing η such that $\mu(S_\eta) > 1 - \varepsilon$ (this is possible since $\lim_{\eta \to +\infty} \mu(S_\eta) = \mu(X) = 1$), we obtain

$$\mu_L(S_\varrho \cap L) = \mu\left(P_L^{-1}(S_\varrho \cap L)\right) \geq \mu(S_\varrho) \geq \mu(S_\eta) > 1 - \varepsilon .$$

The proof of *sufficiency* is more difficult. We define on the algebra $\mathfrak{B}_0 = \bigcup_L \mathfrak{B}^L$ a finitely additive function μ be means of

$$\mu(A) = \mu_L(A), \quad A \subset \mathfrak{B}^L .$$

To convince ourselves that μ can be extended to a measure defined on $(X, \mathfrak{B})$, it is sufficient to show that μ is continuous on $\mathfrak{B}_0$, i.e., that for an arbitrary sequence of sets $A_n \in \mathfrak{B}_0$ for which $A_n \supset A_{n+1}$ and $\bigcap_n A_n = \phi$ (ϕ is the empty set),

$$\lim_{n \to \infty} \mu(A_n) = 0 . \tag{2}$$

Let A_n be a cylinder set with base in L_n and $L_n \subset L_{n+1}$. Let $B_n \subset L_n$ be the base of A_n. We remark that it is sufficient to prove (2) merely for sets with closed bases. Indeed, choosing closed sets $C_n \subset B_n$ such that $\mu_{L_n}(B_n - C_n) < \varepsilon_n$ and then taking

$$D_n = \bigcap_{m=1}^{n} \{x: P_{L_m} x \in C_m\} \cap L_n ,$$

we obtain closed sets for which

$$\mu_{L_n}(B_n - D_n) \leq \sum_{m=1}^{n} \mu_{L_m}(B_m - C_m) \leq \sum_{m=1}^{n} \varepsilon_m$$

Hence, if $A_n' = P_{L_n}^{-1}(D_n)$, then

$$A_{n+1}' \subset A_n', \; \bigcap_n A_n' = \bigcap_n A_n \qquad \text{and} \qquad \mu(A_n) \leq \mu(A_n') + \sum_{m=1}^{n} \varepsilon_m$$

If for cylinder sets with closed bases $\lim_{n\to\infty} \mu(A_n') = 0$, then since $\sum_{m=1}^{\infty} \varepsilon_n$ can be taken as suitably small, (2) is also fulfilled. Hence, it will be assumed that the B_n are closed sets. Then the A_n will be weakly closed sets (if $x_k \xrightarrow{\text{w.}} x$ and $x_k \in A_n$, then $x \in A_n$). For all ϱ the set S_ϱ is also weakly closed and weakly compact. Since

$$S_\varrho \cap \left[\bigcap_{n=1}^{\infty} A_n\right] = \phi ,$$

$\bigcup_{n=1}^{\infty} [S_\varrho \cap A_n] = \phi$ and since the sets $S_\varrho \cap A_n$ are weakly closed and weakly compact and $S_\varrho \cap A_n \supset S_\varrho \cap A_{n+1}$, for some n $S_\varrho \cap A_n = \phi$. This means that

$$\mu(A_n) = \mu_{L_n}(A_n) \leq \mu_{L_n}(L_n) - \mu_{L_n}(L_n \cap S_\varrho) \leq \varepsilon ,$$

provided that $\varrho > \eta$ (η and ε are defined in the lemma). From the arbitrariness of $\varepsilon > 0$ there follows (2). □

Remark. Let L_n be a sequence of finite-dimensional subspaces for which $L_n \subset L_{n+1}$ and $\bigcup L_n$ is dense in X, and μ_{L_n} a sequence of finite-dimensional distributions. This sequence generates some measure iff for arbitrary $\varepsilon > 0$ there exists an $\eta > 0$ such that for all n $\mu_{L_n}(S_\varrho \cap L_n) \geq \geq 1 - \varepsilon$ when $\varrho > \eta$. The proof of this fact is carried out as in the proof of Lemma 1.

It is interesting to note that some classes of functions can also be integrated w.r.t. a weak distribution. To these functions, for example, are related the "cylinder functions" defined below. The function $\varphi(x)$ is called a *cylinder function* if for some finite-dimensional subspace L it is $\mathfrak{B}^L$-measurable. In other words, every cylinder function $\varphi(x)$ has the form

$$\varphi(x) = \varphi_L(P_L x) , \tag{3}$$

where φ_L is some $\mathfrak{B}_L$-measurable function defined on L, and L is a finite-dimensional subspace of X. For each nonnegative cylinder function $\varphi(x)$ we define the "integral" w.r.t. the weak distribution $\{\mu_L\}$ which will be denoted by μ_* to distinguish it from a measure. This integral is defined by the relation

$$\int \varphi(x)\, \mu_*(dx) = \int \varphi_L(x)\, \mu_L(dx) , \tag{4}$$

where φ_L is as in the representation (3). Since (3) is not unique, we must show that

$$\int \varphi_L(x)\, \mu_L(dx)$$

does not depend on the choice of L. Let $L_1 \subset L_2$ and

$$\varphi(x) = \varphi_{L_2}(P_{L_2} x) = \varphi_{L_1}(P_{L_1} x) .$$

Then for $x \in L_2$

$$\varphi_{L_1}(P_{L_1} x) = \varphi_{L_2}(x) .$$

Hence,

$$\int_{L_2} \varphi_{L_2}(x) \, \mu_{L_2}(dx) = \int_{L_2} \varphi_{L_1}(P_{L_2} x) \, \mu_{L_2}(dx) =$$
$$= \int_{L_2} \varphi_{L_1}(y) \, \mu_{L_2}(P_{L_1}^{-1}(dy)) = \int_{L_1} \varphi_{L_1}(x) \, \mu_{L_1}(dx)$$

because of the compatibility condition. This implies that the left side in (4) is uniquely defined.

Denote by B_+ the set of nonnegative $\mathfrak{B}$-measurable functions and by C_+ the set of nonnegative cylinder functions. The symbol $\tilde{B}_+$ will denote those functions $\varphi \in B_+$ for which $\varphi(x) = \lim \psi_n(x)$, where $\psi_n \in C_n$ and $\psi_n \geq \varphi$. If $\varphi \in \tilde{B}_+$, then we put

$$\int \varphi(x) \, \mu_*(dx) = \inf_{\psi \geq \varphi} \int_{\psi \in C_+} \psi(x) \, \mu_*(dx) .$$

Obviously, for $\varphi_1, \varphi_2 \in B_+$

$$\int (\varphi_1(x) + \varphi_2(x)) \, \mu_*(dx) \leq \int \varphi_1(x) \, \mu_*(dx) + \int \varphi_2(x) \, \mu_*(dx) . \quad (5)$$

It is possible to define classes of functions for which equality always holds in (5). One of these will be the class of nonnegative cylinder functions. Another, the class of functions which can be represented by sums of uniformly convergent series of nonnegative cylinder functions. Let K_+ be some class of functions from $\tilde{B}_+$ for which: a) if φ_1 und $\varphi_2 \in K_+$ then also $\varphi_1 + \varphi_2 \in K_+$, b) if φ_1 and $\varphi_2 \in K_+$ then equality holds in (5). If K is the set of functions of the form $\varphi_1 - \varphi_2$, where $\varphi_1, \varphi_2 \in K^+$, then we can define

$$\int (\varphi_1(x) - \varphi_2(x)) \, \mu_*(dx) = \int \varphi_1(x) \, \mu_*(dx) - \int \varphi_2(x) \, \mu_*(dx) . \quad (6)$$

(if at least one of the integrals on the left is finite). It follows from b) that such a definition is unique. The set K_+ can always be defined by a cone (i.e. a set containing $\lambda \varphi$ for $\lambda > 0$ when it contains φ). Then K will be a linear manifold of functions.

Consider the totality of all cones B_+ satisfying Conditions a) and b). A cone will be called maximal if it is not contained in any other. Each cone is contained in some maximal one. The intersection of all maximal cones will be denoted by B_+^0. It is not empty since it contains the cone of nonnegative cylinder functions. Functions from B_+^0 are called integrable w.r.t. the weak distribution μ_*. Let us describe the class B_+^0. For any function $\psi \in \tilde{B}_+$, the set $K_+ = \{\varphi : \varphi = \lambda \psi + \varphi_1, \lambda \geq 0, \varphi_1 \in C_+\}$ is a cone satisfying Conditions a) and b) and containing ψ. Hence, every function $\psi \in \tilde{B}_+$ belongs to some cone. This means that for arbitrary functions $\psi \in \tilde{B}_+$ and $\varphi \in B_+^0$ we have

$$\int (\psi(x) + \varphi(x))\, \mu_*(dx) = \int \psi(x)\, \mu_*(dx) + \int \varphi(x)\, \mu_*(dx)\,. \qquad (7)$$

It is easy to verify the converse: if φ satisfies (7) for all $\psi \in B_+^0$, then $\varphi \in B_+^0$. Thus, the class B_+^0 is completely characterized.

We note that when the weak distribution μ_* is generated by the measure μ, then

$$\int \varphi(x)\, \mu(dx) = \int \varphi(x)\, \mu_*(dx)$$

on B_+^0 and $\tilde{B}_+$. This fact can be used to construct an example of a weak distribution which does not possess a corresponding measure.

Example. Let $m_L(dx)$ denote Lebesgue measure on the subspace L and set $\alpha_L = (2\pi)^{-n}$, where $2n$ is the dimension of L. Finally, put

$$\mu_L(A) = \alpha_L \int\limits_A \exp\left\{-\frac{1}{2}(x, x)\right\} m_L(dx)\,.$$

The compatibility conditions follow from

$$\alpha_L \int\limits_X \exp\left\{-\frac{1}{2}(x, x)\right\} m_L(dx) = 1$$

since, if $L_2 = L_1 + L$, where L is orthogonal to L_1, then

$$\mu_{L_2}(P_{L_1}^{-1}(A) \cap L_2) = \alpha_{L_2} \int\limits_{x \in P_{L_1}^{-1}(A) \cap L_2} \exp\left\{-\frac{1}{2}(x, x)\right\} m_{L_2}(dx) =$$

$$= \alpha_{L_1} \cdot \alpha_L \int\limits_A \exp\left\{-\frac{1}{2}(x_1, x_1)\right\} m_{L_1}(dx) \cdot \int\limits_L \exp\left\{-\frac{1}{2}(x, x)\right\} m_L(dx) =$$

$$= \mu_{L_1}(A)\,.$$

Note that for the cylinder function

$$\varphi_L(x) = \exp\{-\varepsilon(P_L x, P_L x)\}$$

one has

$$\int \varphi(x)\, \mu_*(\mathrm{d}x) = \alpha_L \int \exp\left\{-\frac{1}{2}(1 + 2\varepsilon)(x, x)\right\} m_L(dx) =$$

$$= \alpha_L(1 + 2\varepsilon)^{-n} \int \exp\left\{-\frac{1}{2}(x, x)\right\} m_L(dx) = (1 + 2\varepsilon)^{-n}\,,$$

where $2n$ is the dimension of the space L. Let L_n be an arbitrary increasing sequence of subspaces such that the dimension of L_n is n and $\bigcup L_n$ is dense in X. For some $\varepsilon > 0$, put

$$\varphi_n(x) = \exp\{-\varepsilon(P_n x, P_n x)\}\,,$$

where P_n is projection onto L_n. Then

$$\int \varphi_n(x)\, \mu_*(dx) = (1 + 2\varepsilon)^{-n/2}\,.$$

Since $\varphi_n(x) \downarrow \varphi(x)$, where $\varphi(x) = \exp\{-\varepsilon(x, x)\}$, and this holds for an arbitrary increasing sequence of subspaces, while for each cylinder function $\psi(x) = \psi_L(P_L x)$ for which $\psi(x) \geq \varphi(x)$ we also have $\psi(x) \geq \geq \varphi_n(x)$ for $L_n \supset L$, we obtain

$$\int \varphi(x)\, \mu_*(dx) = \lim_{n\to\infty} \int \varphi_n(x)\, \mu_*(dx) = \lim_{n\to\infty} (1 + 2\varepsilon)^{-n/2} = 0\,.$$

If there existed a measure μ for which μ_L were the set of finite-dimensional distributions, then we would have

$$\int \varphi(x)\, \mu(dx) = \int \varphi(x)\, \mu_*(dx) = 0\,.$$

This equality is impossible since φ is everywhere positive.

Certain facts manifested in the example constructed above can be used to formulate another condition for the existence of a measure with given finite-dimensional distributions.

Lemma 2. *In order that the weak distribution $\{\mu_L\}$ be generated by some measure μ on $(X, \mathfrak{B})$ it is necessary and sufficient that*

$$\lim_{\varepsilon \downarrow 0} \int \exp\{-\varepsilon(x, x)\}\, \mu_*(dx) = 1\,. \tag{8}$$

Proof. In the example considered above it was proved that the function $\exp\{-\varepsilon(x, x)\}$ belongs to $\tilde{B}_+$. Hence, the integral

$$\int \exp\{-\varepsilon(x, x)\}\, \mu_*(dx)$$

makes sense and coincides with

$$\int \exp\{-\varepsilon(x, x)\}\, \mu(dx)$$

when the weak distribution μ_* is generated by the measure μ. Since

$$\lim_{\varepsilon \downarrow 0} \int \exp\{-\varepsilon(x, x)\}\, \mu(dx) = \int \mu(dx) = \mu(X) = 1\,,$$

the necessity of (8) follows. Now assume (8) holds. Then for each $\delta > 0$ one can choose an $\varepsilon > 0$ such that

$$\int \exp\{-\varepsilon(x, x)\}\, \mu_*(dx) > 1 - \delta\,.$$

Hence, for all finite-dimensional subspaces L

$$1 - \delta < \int \exp\{-\varepsilon(P_L x, P_L x)\}\, \mu_*(dx) =$$
$$= \int_L \exp\{-\varepsilon(x, x)\, \mu_L(dx) \leq \int_{L \cap S_\varrho} \mu_L(dx) + e^{-\varepsilon\varrho^2}\left[1 - \int_{L \cap S_\varrho} \mu_L(dx)\right] =$$
$$= e^{-\varepsilon\varrho^2} + (1 - e^{-\varepsilon\varrho^2})\, \mu_L(S_\varrho \cap L)\,.$$

Thus,

$$\mu_L(S_\varrho \cap L) \geq 1 - \delta\, (1 - e^{-\varepsilon\varrho^2})\,.$$

Choosing first δ and then $\eta = \varepsilon^{-1/2}$ we find that for $\varrho > \eta$

$$\mu_L(S_\varrho \cap L) \geq 1 - \delta\,(1 - e^{-1})\,.$$

Now apply Lemma 1. □

We note an important property of measures in Hilbert space.

Theorem 1. *For any finite measure μ on $(X, \mathfrak{B})$ and for each $\varepsilon > 0$ there exists a compact $K_\varepsilon \subset X$ such that $\mu\,(X - K_\varepsilon) < \varepsilon$.*

Proof. Let N be some countable everywhere dense subset of X and $S_\varrho(x)$ the sphere with center at x and radius ϱ. Since for all ϱ

$$X = \bigcup_{x \in N} S_\varrho(x)\,,$$

for any ϱ and η we can find a finite set of points $x_1, \ldots, x_n \in N$ such that

$$\mu\left(X - \bigcup_{k=1}^{n} S_\varrho(x_k)\right) < \eta\,.$$

Let $x_1, \ldots, x_{l_n}$ be a set of points from N for which

$$\mu\left(X - \bigcup_{k=1}^{l_n} S_{1/n}(x_k)\right) < \frac{\varepsilon}{2}\,n\,.$$

Denote by D_n the family of sets $\bigcup_{k=1}^{l_n} S_{1/n}(x_k)$. Then $\bigcap_{n=1}^{\infty} D_n = D$ is closed and

$$\mu\,(X - D) \leq \sum_{n=1}^{\infty} \mu\,(X - D_n) < \sum_{n=1}^{\infty} \varepsilon \cdot \frac{1}{2^n} = \varepsilon\,.$$

Finally note that D is a compact since for $1/n < \delta$ the points $x_1, \ldots, x_{l_n}$ form a δ-net in D_n and hence also in D. □

Corollary. For any $\mathfrak{B}$-measurable set B one can find a compact K for which $\mu\,(B - K) < \varepsilon$. From Theorem 1 it follows that this holds for all closed sets B. We show that for all measurable sets B one can find for each $\varepsilon > 0$ a closed set F such that $F \subset B$ and $\mu\,(B - F) < \varepsilon$. Let $\mathfrak{A}$ be the class of such sets. Clearly, when $\mathfrak{A}$ contains $B_1, \ldots, B_n, \ldots$ then it also contains

$$\bigcup_{n=1}^{\infty} B_n \quad \text{and} \quad \bigcap_{n=1}^{\infty} B_n\,.$$

$\mathfrak{A}$ also contains all open and closed sets, which means that it contains the algebra $\mathfrak{A}_0$ generated by closed and open sets; moreover, $\mathfrak{A}$ is a monotone class. Hence, it follows that $\mathfrak{A}$ coincides with $\mathfrak{B}$.

§ 3. The Characteristic Functional. Moment Functionals

Assume a normalized measure μ is given on $(X, \mathfrak{B})$. The complex-valued function $\exp\{i(z, x)\}$ (z fixed) is bounded and $\mathfrak{B}$-measurable. Hence, for all $z \in X$ the integral

$$\theta(z) = \int \exp\{i(z, x)\}\, \mu(dx) \tag{1}$$

is defined. The function $\theta(z)$, defined on X, satisfies the following conditions: 1) $\theta(0)=1$, 2) $\theta(z)$ is continuous (even weakly[1] continuous) in z, 3) it is positive definite, i.e., for an arbitrary set of $z_1, \ldots, z_N$

$$\sum_{l,k=1}^{N} \theta(z_l - z_k)\, \alpha_l \bar{\alpha}_k \geq 0$$

for all complex $\alpha_1, \ldots, \alpha_N$. The latter follows from the equality

$$\sum_{l,k=1}^{N} \theta(z_l - z_k)\, \alpha_l \bar{\alpha}_k = \int \left| \sum_{k=1}^{N} \alpha_k \exp\{i(z_k, x)\} \right|^2 \mu(dx)\,.$$

The function $\theta(z)$ is called the *characteristic functional* of the measure μ. It is distinguished by the fact that it uniquely determines this measure. To show this, we note that for $z \in L$, where L is some finite-dimensional subspace of X, we have

$$\int \exp\{i(z, x)\}\, \mu(dx) = \int \exp\{i(z, P_L x)\}\, \mu(dx) = \int \exp\{i(z, x)\}\, \mu_L(dx)$$

Hence, considering $\theta(z)$ only for $z \in L$, we obtain the characteristic function (Fourier transformation) of the measure μ_L, which is uniquely determined by its Fourier transformation. This means that the characteristic functional of a measure determines all finite-dimensional distributions $\{\mu_L\}$ which, in turn, determine the measure.

We note that for the definition of $\theta(z)$ it is sufficient to know μ_L merely for one-dimensional L. Indeed, let L_z be a one-dimensional subspace spanned by the vector z. Then

$$\int \exp\{i(z, x)\}\, \mu(dx) = \int \exp\{i(z, P_{L_z} x)\}\, \mu(dx) = \\ = \int \exp\{i(z, x)\}\, \mu_{L_z}(dx)\,.$$

Thus, the measure μ is defined by its one-dimensional distributions (i.e. by its projections onto one-dimensional subspaces).

Now let a function $\theta(z)$ be given which satisfies Conditions 1)—3). Then on each finite-dimensional subspace L this function will be positive-definite and continuous so that by a theorem of Bochner, $\theta(z)$ is representable for $z \in L$ in the form

$$\theta(z) = \int \exp\{i(z, x)\}\, \mu_L(\mathrm{d}x)\,, \tag{2}$$

where μ_L is some measure on $(L, \mathfrak{B}_L)$ for which

$$\mu_L(L) = \theta(0) = 1$$

[1] It is continuous w.r.t. the weak topology in X.—Transl.

because of Condition 1). We now show that the measures μ_{L_1} and μ_{L_2} are compatible for $L_1 \subset L_2$. It suffices to show that for any continuous bounded function $\varphi(x)$ on L_1

$$\int_{L_2} \varphi(P_{L_1} x)\, \mu_{L_2}(dx) = \int_{L_1} \varphi(x)\, \mu_{L_1}(dx) . \tag{3}$$

First let

$$\varphi_z(x) = \exp\{i(z, x)\}, \qquad z, x \in L_1 .$$

Then

$$\int_{L_2} \exp\{i(z, P_{L_1} x)\}\, \mu_{L_2}(dx) = \int_{L_2} \exp\{i(z, x)\}\, \mu_{L_2}(dx) =$$
$$= \theta(z) = \int_{L_1} \exp\{i(z, x)\}\, \mu_{L_1}(dx) .$$

This means that (2) is fulfilled for $\varphi_z(x)$. It remains to note that it is possible to construct a sequence of linear combinations of the form

$$\sum_{k=1}^{N} \alpha_k\, \varphi_{z_k}(x)$$

all of which are bounded and which converge everywhere to $\varphi(x)$.

Hence, to any functional $\theta(z)$ satisfying Conditions (1)—(3) there corresponds a compatible family of finite-dimensional distributions $\{\mu_L\}$ in such a way that (2) is satisfied. In other words, to each such functional there corresponds a weak distribution. Since $\exp\{i(z, x)\}$ is a cylinder function, one can define the function

$$\theta(z) = \int \exp\{i(z, x)\}\, \mu_*(dx)$$

for any weak distribution μ_*. Such a functional will be called the *characteristic functional* for the weak distribution. It satisfies Conditions (1) and (3) and instead of Condition (2) it fulfills: (2[1]) $\theta(z)$ is continuous in z in any finite-dimensional subspace $L \subset X$ (i.e., a bounded $\theta(z)$ is continuous on this subspace).

Characteristic functionals of measures and weak distributions are of use in giving compatible families of finite-dimensional distributions. Of course, not every functional satisfying (1)—(3) can be associated with a measure. Additional conditions which must be imposed on functionals in order that this be possible will be set up in the next section. We will also give examples which show that Conditions (1)—(3) are not sufficient for the existence of a corresponding measure.

We now treat some other characteristics of a measure. Assume that for all $z_1, \ldots, z_N \in X$ the integral

$$\int (z_1, x) \cdots (z_N, x)\, \mu(dx) = \sigma_N(z_1, \ldots, z_N)$$

exists. Then the function $\sigma_N(z_1, \ldots, z_N)$ is called the *moment function of order* N of the measure μ. It is easy to see that the moment function is symmetric, additive and homogeneous in each argument:

$$\sigma_N(\lambda_1 z_1' + \lambda_2 z_1'', \ldots, z_N) = \lambda_1 \sigma_N(z_1', \ldots, z_N) + \lambda_2 \sigma_N(z_1'', \ldots, z_N) \quad (4)$$

(symmetry implies the additivity and homogeneity in the remaining arguments).

Theorem 1. *If the N-th moment function is defined for the measure μ, then it is a bounded N-linear form.*

Proof. From (4) and the symmetry of $\sigma_N(z_1, \ldots, z_N)$ it follows that it suffices to demonstrate the boundedness of σ_N, i.e., the existence of a constant α_N such that

$$|\sigma_N(z_1, \ldots, z_N)| \le \alpha_N |z_1| \cdots |z_N| . \quad (5)$$

For (5) to hold it is necessary and sufficient that

$$\sup_{|z| \le 1} \int |(z, x)|^N \mu(dx) < \infty . \quad (6)$$

Introduce the functions

$$\sigma_N^{(k)}(z) = \int \frac{k\,|(z, x)|^N}{k + |x|^N} \mu(dx) , \quad \bar{\sigma}_N(z) = \int |(z, x)|^N \mu(dx) .$$

For arbitrary k, $\sigma_N^{(k)}(z)$ is weakly continuous in z. Furthermore, for all $z \in X$ $\sigma_N^{(k)}(z) \uparrow \bar{\sigma}_N(z)$ for $k \uparrow \infty$. Set

$$K_l^{(k)} = \{z : \sigma_N^{(k)}(z) \ge l\} \cap \{z : |z| \le 1\} .$$

The set $K_l^{(k)}$ is weakly closed due to the weak continuity of the function $\sigma_N^{(k)}$ and is weakly compact since it is bounded. Obviously,

$$\bigcap_k K_l^{(k)} = \{z : \bar{\sigma}_N(z) \ge l\} .$$

Hence, for the proof of (5) it is sufficient to show that $\bigcap_k K_l^{(k)}$ is empty for some l. This set is also weakly compact and weakly closed. If for all natural l the set $\bigcap_k K_l^{(k)}$ were not empty, then $\bigcap_{l=1}^{\infty} \bigcap_k K_l^{(k)}$ would also not be empty which is impossible since for $z \in \bigcap_l \bigcap_k K_l^{(k)}$ we would then have $\bar{\sigma}_N(z) = +\infty$. □

Moment functions can be expressed by means of the characteristic functional. For the functional $\theta(z)$ defined on X let

$$\theta^{(N)}(z; z_1, \ldots, z_N) = \frac{\partial^N}{\partial\lambda_1 \cdots \partial\lambda_N} \theta\left(z + \sum_1^N \lambda_k z_k\right)\Bigg|_{\lambda_1 = \cdots = \lambda_N = 0} \quad (7)$$

which we will call the N-th weak derivative (it is an N-linear form in $z_1, \ldots, z_N$). Then, if $\theta(z)$ is the characteristic functional of the measure

μ and the N-th-order moment function σ_N exists, we have

$$i^N \sigma_N(z_1, \ldots, z_N) = \theta^{(N)}(0; z_1, \ldots, z_N) . \tag{8}$$

To obtain (8) we need only apply (7) to (1). When N is even, it follows from the existence of the N-th weak derivative of $\theta(z)$ that σ_N exists and we again have (8). It is easy to show this starting from (1) and calculating $\theta^N(0; z_1, \ldots, z_N)$. One uses induction on N and the fact that the second derivative of a function of a real-valued parameter $\alpha(\lambda)$ is

$$\alpha''(0) = \lim_{\lambda \to 0} \frac{1}{\lambda^2} [\alpha(\lambda) + \alpha(-\lambda) - 2\alpha(0)] .$$

The first two moment functions find the widest application. If $\sigma_1(z)$ is defined, it is linear functional on X and consequently, there exists a vector a such that

$$\sigma_1(z) = (a, z) .$$

This vector a is called the *mean value* for the measure μ. For a we have for all $z \in X$

$$(a, z) = \int (x, z)\, \mu(dx) . \tag{9}$$

The second moment function $\sigma_2(z_1, z_2)$ is a bilinear functional. This means that there exists a bounded symmetric linear operator A_1 on X such that

$$\sigma_2(z_1, z_2) = (A_1 z_1, z_2) .$$

This operator is called the *covariance operator* of the measure μ. It is defined by the relation

$$(A_1 z_1, z_2) = \int (x, z_1)(x, z_2)\, \mu(dx) . \tag{10}$$

More useful is the *correlation operator*, defined by

$$(A z_1, z_2) = \sigma_2(z_1, z_2) - \sigma_1(z_1)\sigma_1(z_2) = \int (x - a, z_1)(x - a, z_2)\, \mu(dx) . \tag{11}$$

It is easy to see that the covariance and correlation operators are nonnegative. The following result will be useful in the sequel.

Lemma 1. *If the measure μ is such that*

$$\int |x|^2 \mu(dx) < \infty ,$$

then the covariance operator A_1 is nuclear and

$$\operatorname{tr} A_1 = \int |x|^2 \mu(dx) .$$

(Here, tr A_1 *denotes the trace of the operator and an operator is nuclear if it is symmetric, nonnegative and has finite trace).*

Proof. Let $\{e_k\}$ be an arbitrary orthonormalized basis in X. Then

$$\sum_{k=1}^{N} (A_1 e_k, e_k) = \int \sum_{k=1}^{N} (x, e_k)^2 \mu(dx) .$$

Letting $N \to \infty$ in this equality we finish the proof. □

Since for the calculation of moment functions it is necessary to integrate cylinder functions, these functions can also be defined for a weak distribution; however, it is then no longer possible to guarantee that they will be bounded multilinear functions.

§ 4. The Minlos-Sazonov Theorem

This theorem gives necessary and sufficient conditions that the function $\theta(z)$, defined on X, be the characteristic functional of some measure on $(X, \mathfrak{B})$. It is thus a generalization of Bochner's theorem on positive definite functions in Hilbert space.

Theorem 1. *In order that the complex-valued, continuous positive-definite function $\theta(z)$, defined on X, be the characteristic functional of a finite measure on $(X, \mathfrak{B})$ it is necessary and sufficient that for any $\varepsilon > 0$ one can determine a nuclear operator S_ε for which* $\operatorname{Re}(\theta(0) - \theta(z)) < \varepsilon$ *when* $(S_\varepsilon, z, z) < 1$.

Proof. Necessity. Let $\theta(z)$ be the characteristic functional of the measure μ. Then for all γ

$$\operatorname{Re}(\theta(0) - \theta(z)) = \int (1 - \cos(z, x))\, \mu(dx) \leq$$

$$\leq \int_{|x| \leq \gamma} 2 \sin^2 \frac{(x, z)}{2} \mu(dx) + 2 \int_{|x| > \gamma} \mu(dx) \leq$$

$$\leq \frac{1}{2} \int_{|x| \leq \gamma} (x, z)^2 \mu(dx) + 2 \mu(\{x : |x| > \gamma\}) .$$

Since

$$\int_{|x| \leq \gamma} |x|^2 \mu(dx) < \infty ,$$

the covariance operator B_γ defined by

$$(B_\gamma z_1, z_2) = \int_{|x| < \gamma} (x, z_1)(x, z_2)\, \mu(dx)$$

is nuclear because of Lemma 1, § 3. Let γ be chosen so that

$$\mu(\{x : |x| > \gamma\}) < \frac{\varepsilon}{4} .$$

Then

$$\operatorname{Re}\left(\theta(0)-\theta(z)\right) < \frac{\varepsilon}{2} + \frac{1}{2}(B_\gamma z, z) < \varepsilon,$$

provided that $\frac{1}{2}(B_\gamma z, z) < \frac{\varepsilon}{2}$. Hence, the condition of the theorem is satisfied if $S_\varepsilon = \frac{1}{\varepsilon} B_\gamma$. (The nuclearity of S_ε follows from that of B_γ). The necessity is proved.

Sufficiency. Assume that $\theta(0) = 1$. One can construct a compatible family of finite-dimensional distributions $\{\mu_L\}$ with respect to $\theta(z)$. We now use Lemma 2 § 2. In order to estimate

$$\int e^{-\varepsilon(x,x)}\,\mu_*(dx)$$

we will find a convenient representation for $\int e^{-\varepsilon(x,x)}\,\mu_L(dx)$. Let m_L again be Lebesgue measure on L and n_L the dimension of L. Then, representing the n_L-tuple integral as a product of simple integrals, we get

$$e^{-\varepsilon(x,x)} = \left(2\sqrt{\pi\varepsilon}\right)^{-n_L} \int e^{i(z,x) - \frac{(z,z)}{4\varepsilon}}\, m_L(dz)\,.$$

Thus,

$$\int e^{-\varepsilon(x,x)}\,\mu_L(dx) = \left(2\sqrt{\pi\varepsilon}\right)^{-n_L} \int e^{-\frac{(z,z)}{4\varepsilon}} \int e^{i(z,x)}\,\mu_L(dx)\, m_L(dz) =$$

$$= \left(2\sqrt{\pi\varepsilon}\right)^{-n_L} \int e^{-\frac{(z,z)}{4\varepsilon}}\,\theta(z)\, m_L(dz)\,.$$

This implies that

$$1 - \int e^{-\varepsilon(x,x)}\,\mu_L(dx) = \left(2\sqrt{\pi\varepsilon}\right)^{-n_L} \int e^{-\frac{(z,z)}{4\varepsilon}} \left(1-\theta(z)\right) m_L(dz) =$$

$$= \left(2\sqrt{\pi\varepsilon}\right)^{-n_L} \int e^{-\frac{(z,z)}{4\varepsilon}} \operatorname{Re}\left(1-\theta(z)\right) m_L(dz)\,.$$

Let S_δ be a nuclear operator for which $\operatorname{Re}\left(1-\theta(z)\right) < \delta$ when $(S_\delta z, z) < 1$. Then

$$1 - \theta(z) < \delta + 2(S_\delta z, z)\,,$$

$$1 - \int e^{-\varepsilon(x,x)}\,\mu_L(dx) \le \delta\left(2\sqrt{\pi\varepsilon}\right)^{-n_L} \int e^{-\frac{(z,z)}{4\varepsilon}} \left[\delta + 2(S_\delta z, z)\right] m_L(dz) =$$

$$= \delta + 8\varepsilon \sum_1^{n_L} (S_\delta e_k, e_k)\,,$$

where $\{e_k\}$ is a basis in L. We used the fact that for an arbitrary orthonormalized basis in L

$$(S_\delta z, z) = \sum_{k,j} (S_\delta e_k, e_j)\,(z, e_k)\,(z, e_j)$$

and

$$(2\sqrt{\pi\varepsilon})^{-n_L}\int e^{-\frac{(z,z)}{4\varepsilon}}(z,e_k)(z,e_j)\,m_L(dz) = 8\varepsilon\delta_j^k,$$
$$\left(\delta_j^k = \begin{cases}1 & \text{for } k=j\\ 0 & \text{for } k\neq j\end{cases}\right).$$

Therefore,

$$1-\int e^{-\varepsilon(x,x)}\mu_*(dx)\leq\delta+8\varepsilon\operatorname{tr}S_\varrho\,.$$

Finally, letting $\varepsilon\downarrow 0$ and then $\delta\downarrow 0$, we find that the condition of Lemma 2 § 2 is satisfied. Hence, there exists a measure μ on $(X,\mathfrak{B})$ for which $\{\mu_L\}$ are finite dimensional distributions. Thus, the characteristic functional of this measure will be $\theta(z)$.

Remark. If $\theta(z)$ is the characteristic functional of some normalized measure, then a nuclear operator S can be found such that

$$|\theta(z_1)-\theta(z_2)|\to 0 \qquad \text{for } \big(S(z_1-z_2),(z_1-z_2)\big)\to 0\,.$$

Indeed,

$$|\theta(z_1)-\theta(z_2)|\leq\int|1-e^{i(z_1-z_2,x)}|\,\mu(dx)\leq$$
$$\leq\Big(\int|1-e^{i(z_1-z_2,x)}|^2\mu(dx)\Big)^{1/2}=(2\,[1-\operatorname{Re}\theta(z_1-z_2)])^{1/2},$$

so that it suffices to demonstrate the existence of a nuclear operator S for which $\operatorname{Re}\theta(z)\to 1$ when $(S\,z,z)\to 0$. Assume that $\varepsilon_k\downarrow 0$ and that S_{ε_k} is a nuclear operator for which

$$\operatorname{Re}\big(1-\theta(z)\big)<\varepsilon_k \qquad \text{when} \qquad (S_{\varepsilon_k}z,z)<1\,.$$

Choose a sequence $\lambda_k>0$ for which

$$\sum_k\lambda_k\operatorname{tr}S_{\varepsilon_k}<\infty\,.$$

Then, the operator $S=\sum\lambda_k S_{\varepsilon_k}$ is nuclear (the series converges in norm since $\operatorname{tr}S_{\varepsilon_k}\geq\|S_{\varepsilon_k}\|$). If $(S\,z,z)<\lambda_k$, then $(S_{\varepsilon_k}z,z)<1$ so that $\operatorname{Re}\big(1-\theta(z)\big)<\varepsilon_k$. It is clear that the existence of such an operator S guarantees that the conditions of the Minlos-Sazonov theorem are satisfied for $\theta(z)$.

Example. Let $\theta(z)=\exp\{-(C\,z,z)\}$, where C is a symmetric, nonnegative bounded operator. If $1-\theta(z)<\varepsilon$, then $(C\,z,z)<$ $<\log\frac{1}{1-\varepsilon}$. If this holds when $(S_\varepsilon z,z)<1$, then

$$(C\,z,z)<(S_\varepsilon z,z)\log\frac{1}{1-\varepsilon}\,.$$

Hence, the condition of the Minlos-Sazonov theorem is satisfied for $\theta(z)$ merely under the condition that the operator C be nuclear. At the same time, $\theta(z)$ will be a continuous positive-definite function for all symmetric nonnegative bounded operators C (the positive-definiteness of $\theta(z)$ follows from the positive-definiteness of $\theta(z)$ on finite-dimensional subspaces of X). Taking nuclear C's, we obtain examples of continuous positive-definite functions which are not characteristic functionals of measures. If C is taken as completely continuous and nonnuclear, then $\theta(z)$ will be weakly continuous but will not be the characteristic functional of a measure.

§ 5. Gaussian Measures

Gaussian measures in finite-dimensional spaces have been investigated extensively in the theory of probability. This is perhaps the simplest class of measures for which generalization is possible to Hilbert space (as we will see later, the Lebesgue measure cannot be generalized to Hilbert space). By a *Gaussian measure in a Hilbert space* $(X, \mathfrak{B})$ we understand a measure μ whose characteristic functional has the form

$$\theta(z) = \exp\left\{i(a, z) - \frac{1}{2}(A z, z)\right\}, \tag{1}$$

where $a \in X$ and A is a symmetric bounded nonnegative operator on X. The expression (1) will actually be the characteristic functional of some measure if A is a nuclear operator. This follows from Theorem 1 § 4 and the relation

$$1 - \operatorname{Re}\theta(z) = 1 - \exp\left\{-\frac{1}{2}(A z, z)\right\} + \exp\left\{-\frac{1}{2}(A z, z)\right\} \times \\ \times \left(1 - \cos(a, z)\right),$$

and is proved exactly as in the example of § 4. It is easy to verify that

$$\theta'(0, z) = i(a, z)$$

$$\theta''(0, z_1, z_2) = -\left[(a, z_1)(a, z_2) + (A z_1, z_2)\right].$$

Hence, the vector a is the mean of the measure μ and A is the correlation operator of this measure.

It follows from (1) that all finite-dimensional projections μ_L of the Gaussian measure μ are also Gaussian measures in the corresponding subspace. It turns out that one can define a set of subspaces L, the projections upon which completely determine the measure μ and on which the structure of μ is simplest.

We will call the measure μ *nondegenerate* if the operator A is strictly positive. We first consider the case of a nondegenerate measure. Since A is a nuclear, it is completely continuous (we remark that the complete continuity of the operator A introduced in (1) follows from the weak continuity of $\theta(z)$). Let $\{e_k\}$ be a complete orthonormalized system of eigenvectors of the operator A and λ_k the corresponding eigenvalues. Let L_n be the subspace generated by the vectors $\{e_1, \ldots, e_n\}$. Then for $z = \sum_1^n \zeta_k e_k \in L_n$

$$\theta(z) = \exp\left\{i \sum_1^n \alpha_k \zeta_k - \frac{1}{2} \sum_1^n \lambda_k \zeta_k^2\right\} = \prod_{k=1}^n e^{i\alpha_k \zeta_k - \frac{1}{2}\lambda_k \zeta_k^2},$$

where $\alpha_k = (a, e_k)$. Using the equality

$$e^{i\alpha_k \zeta_k - \frac{1}{2}\lambda_k \zeta_k^2} = \frac{1}{\sqrt{2\pi\lambda_k}} \int e^{i\zeta_k \tau - \frac{(\tau - \alpha_k)^2}{2\lambda_k}} d\tau,$$

we get

$$\theta(z) = \prod_{k=1}^n \frac{1}{\sqrt{2\pi\lambda_k}} \int e^{i\zeta_k \tau_k - \frac{\tau_k^2}{2\lambda_k}} d\tau_k =$$

$$= \prod_{k=1}^n (2\pi\lambda_k)^{-\frac{1}{2}} \int e^{-i(z,x) - \frac{1}{2}(A^{-1}(x-a_n),\, x-a_n)} m_{L_n}(dx).$$

We replace the product of integrals with a single one by setting $x = \sum_1^n \tau_k e_k$, $a_n = \sum_1^n \alpha_k e_k$ and $m_{L_n}(dx) = \prod_1^n d\tau_k$, where m_{L_n} again stands for Lebesgue mesaure on L_n. Hence, for any Borel set $B \in \mathfrak{B}_{L_n}$

$$\mu_{L_n}(B) = \left(\prod_{k=1}^n (2\pi\lambda_k)^{-1/2}\right) \int \exp \times$$

$$\times \left\{-\frac{1}{2}(A^{-1}(x - a_n),\, x - a_n)\right\} m_{L_n}(dx). \tag{2}$$

Let $L^{(k)}$ be a subspace spanned by e_k. Then analogously to (2), we find that

$$\mu_{L^{(k)}}(B) = (2\pi\lambda_k)^{-1/2} \int_B \exp\left\{-\frac{1}{2}(A^{-1}(x - \alpha_k e_k),\, x - \alpha_k e_k)\right\} m_{L^{(k)}}(dx).$$

Since $L_n = L^{(1)} \times \cdots \times L^{(n)}$ (more precisely, L_n may be considered as this product), we have from (2) that

$$\mu_{L_n} = \prod_{k=1}^n \mu_{L^{(k)}}.$$

We remark that X itself can be viewed as a subset of the infinite Cartesian product $L_\infty = \prod_{k=1}^\infty L^{(k)}$. One may consider the measure $\prod_{k=1}^\infty \mu_{L^{(k)}}$ in this Cartesian product. Let $x_k \in L^{(k)}$ and $(x_1, x_2, \ldots)$ be a point in L_∞.

According to the definition of an infinite product of measures, $\prod_{k=1}^{\infty} \mu_{L(k)}$ is defined on sets C of the form

$$C = \{(x_1, x_2, \ldots): x_k \in B_k,\ B_k \in \mathfrak{B}_{L(k)},\ k = 1, \ldots, N\}$$

(these sets are called *cylinder sets* in L_∞) by means of the equality

$$\left(\prod_{k=1}^{\infty} \mu_{L(k)}\right)(C) = \prod_{k=1}^{n} \mu_{L(k)}(B_k) = \mu(\{x: P_{L(k)}\, x \in B_k,\ k = 1, \ldots, n\})\,.$$

Then the measure $\left(\prod_{k=1}^{\infty} \mu_{L(k)}\right)(C)$ is extended to the minimal σ-algebra $\mathfrak{B}_{L_\infty}$ containing all cylinder sets in L_∞. On the other hand, the measure μ can also be extended to L_∞ by setting for each measurable set $C \in \mathfrak{B}_{L_\infty}$

$$\mu_{L_\infty}(C) = \mu\,(C \cap X)\,.$$

Since on cylinder sets from L_∞ the measures $\prod_{k=1}^{\infty} \mu_{Lk}$ and μ_{L_∞} coincide, they coincide on $\mathfrak{B}_{L_\infty}$. Hence,

$$\mu_{L_\infty} = \prod_{k=1}^{\infty} \mu_{L(k)}\,.$$

But μ_{L_∞}, defined on $(L_\infty, \mathfrak{B}_{L_\infty})$, has in fact X as its support, since X, as is easily seen, is measurable w.r.t. $\mathfrak{B}_{L_\infty}$:

$$X = \bigcup_{k=1}^{\infty} \bigcap_{n=1}^{\infty} C_{n,k}\,,$$

where $C_{n,k}$ is a cylinder set from L_∞ of the form

$$C_{n,k} = \left\{(x^1, x^2, \ldots): \sum_{1}^{n} (x^i, e_i)^2 \leq k,\ x^i \in L_i\right\}.$$

This means that the measures μ_{L_∞} and μ can be identified with one another so that it makes sense to write

$$\mu = \prod_{k=1}^{\infty} \mu_{L(k)}\,. \tag{3}$$

Thus, the nondegenerate Gaussian measure defined in the sense given above is an infinite product of measures defined on an orthogonal system of linear one-dimensional subspaces. Measures admitting such a representation are sometimes called *product measures*.

Now let μ be a degenerate Gaussian measure. The symbol X_1 will denote the closure of the range of the operator A and X_2 its orthogonal complement. We will view X as the Cartesian product $X_1 \times X_2$. Consider the projections of the measure μ on the subspaces X_1 and X_2 (the projection on an infinte-dimensional subspace is constructed just as in the case of a finite-dimensional one): μ^1 and μ^2. The characteristic functionals of these measures $\theta_k(z)$ are obtained from $\theta(z)$ if z is taken in the corresponding subspace X_k:

$$\theta_1(z) = \exp\left\{i(a_1, z) - \frac{1}{2}(A\,z, z)\right\}, \quad z \in X_1, \quad a_1 = P_{X_1}\, a\,,$$

$$\theta_2(z) = \exp\{i(a_2, z)\}\,, \qquad z \in X_2, \quad a_2 = P_{X_2}\, a\,,$$

where P_{X_k} is projection on X_k. Note that for $z_k \in X_k$

$$\theta(z_1 + z_2) = \theta(z_1)\,\theta(z_2)\,. \tag{4}$$

We now show that $\mu = \mu_1 \times \mu_2$ follows from (4). To do this it is sufficient to show that for any pair of continuous bounded cylinder functions $\varphi_1(x_1)$ and $\varphi_2(x_2)$ (since such functions can approximate the indicators of open cylinder sets in the sense of bounded pointwise convergence) defined resp. on X_1 and X_2, we have

$$\int \varphi(x)\,\mu(dx) = \int \varphi_1(x_1)\,\mu_1(dx_1) \cdot \int \varphi_2(x_2)\,\mu_2(dx_2)\,, \tag{5}$$

where $\varphi(x) = \varphi(x_1)\,\varphi(x_2)$ if $x = x_1 + x_2$. From (4) follows the validity of (5) for the functions

$$\varphi_k(x_k) = \exp\{i(z_k, x_k)\}\,, \qquad z_k \in X_k\,,$$

so that (5) also holds for functions of the form

$$\varphi_k(x_k) = \sum \alpha_{j,k} \exp\{i(z_k^j, x_k)\}\,, \qquad z_k^j \in X_k\,.$$

By means of such functions one can approximate an arbitrary bounded continuous cylinder function in the sense of bounded pointwise convergence so that (5) also holds for these. Its easy to see that (4) holds when $\mu = \mu_1 \times \mu_2$. We have thus proved

Lemma 1. *Let X_1 and X_2 be mutually orthogonal subspaces of X such that $X = X_1 + X_2$ and let μ be a measure on $(X, \mathfrak{B})$. Let μ_k be the projection of μ on X_k and $\theta(z)$, $z \in X$, $\theta(z_k)$, $z_k \in X_k$ characteristic functionals of the measures μ and μ_k, resp. Then in order that $\mu = \mu_1 \times \mu_2$ it is necessary and sufficient that* (4) *be satisfied.*

We return to the Gaussian measure. Obviously, μ_1 will be a nondegenerate Gaussian measure in X_1. Hence, a representation similar to (3) holds for it (only in X_1). Concerning the measure μ_2 in X_2, it follows from the form of $\theta_2(z)$ that this measure is concentrated at the point a_2:

$$\mu_2(B) = 1 \quad \text{if} \quad a_2 \in B,\ \mu_2(B) = 0 \quad \text{if} \quad a_2 \notin B,\ B \in \mathfrak{B}_{X_2}\,.$$

It is clear that μ_2 can also be represented in the form (3) by choosing an arbitrary sequence of orthogonal one-dimensional subspaces L_n'' in X_2. If $\sum L_n''$ is dense in X_2, then denoting by a_n'' the projection of a_2 on L_n''

and by μ_n'' the measure on $\mathfrak{B}_{L_n''}$ for which $\mu_n''(B) = 1$ for $a_n'' \in B$; $\mu_n''(B) = 0$ for $a_n'' \notin B$, $B \in \mathfrak{B}_{L_n''}$, we obtain

$$\mu_2 = \Pi \mu_n'' .$$

Since each of the measures μ_1 and μ_2 is a product measure, so also will be μ.

The representation of a measure as a product of measures on one-dimensional subspaces is quite convenient in the calculation of integrals w.r.t. these measures. Let us evaluate an integral which will be of use later.

Example. Let μ be a Gaussian measure with characteristic functional (1). We will calculate

$$\int \exp\{-\varepsilon(x, x) + (z, x)\} \mu(dx) . \tag{6}$$

Choosing an orthonormalized basis $\{e_k\}$ of eigenvectors of the operator A and denoting by $L^{(k)}$ the subspace spanned by e_k; by x_k, z_k and a_k the projections of x, z and a on $L^{(k)}$, resp., and by λ_k the eigenvalues of the operator A corresponding to the vectors e_k (λ_k can equal 0), we represent (6) as

$$\prod_{k=1}^{\infty} \int_{L^{(k)}} \exp\{-\varepsilon(x_k, x_k) + (z_k, x_k)\} \mu_{L^{(k)}}(dx) . \tag{7}$$

Each factor in the product (7) is easy to calculate. If $\lambda_k > 0$, then

$$\int \exp\{-\varepsilon(x_k, x_k) + (z_k, x_k)\} \mu_{L^{(k)}}(dx) =$$

$$= \frac{1}{\sqrt{2\pi\lambda_k}} \int_{-\infty}^{\infty} \exp\left\{-\varepsilon\tau^2 + (z_k, e_k)\tau - \frac{(\tau - (a, e_k))^2}{2\lambda_k}\right\} d\tau =$$

$$= \frac{1}{\sqrt{1 + 2\varepsilon\lambda_k}} \exp\left\{\frac{1}{2} \cdot \frac{\lambda_k(z, e_k)^2 + 2(a, e_k)(z, e_k) - 4\varepsilon(a, e_k)^2}{1 + 2\varepsilon\lambda_k}\right\} .$$

It is also easy to verify that this also remains valid when $\lambda_k = 0$. Substituting this into (7) and taking into account the connection between λ_k, e_k and A, we obtain

$$\int \exp\{-\varepsilon(x, x) + (z, x)\} \mu(dx =$$

$$= \prod_{k=1}^{\infty} (1 + 2\varepsilon\lambda_k)^{-1/2} \exp\left\{\frac{1}{2}[(A(I + 2\varepsilon A)^{-1} z, z) + \right.$$

$$\left. + ((I + 2\varepsilon A)^{-1} a, z + 2\varepsilon a)]\right\} , \tag{8}$$

where I is the identity operator in X.

§ 6. Generalized Measures in Hilbert Space

In § 2 we associated with each compatible family of finite-dimensional distributions an object called a weak distribution in X. Under the conditions of the Minlos-Sazonov theorem, to a weak distribution there corresponds some measure on $(X, \mathfrak{B})$, i.e., it is generated by some measure. We will say, whenever there is no such measure on $(X, \mathfrak{B})$, that the weak distribution is generated by a generalized measure. Analyzing Lemma 1 § 2, we convince ourselves that a measure corresponding to the weak distribution is lacking because it cannot be "accommodated" in X. Is it possible to define it reasonably on some extension of X? It will turn out that the answer to this question is affirmative. A suitable extension of the Hilbert space can be obtained by using a construction from the theory of equipped Hilbert spaces. It turns out that extensions obtained in such a way can accommodate measures generated by all possible weak distributions on the original Hilbert space. Such measures on extensions of the initial space (satisfying certain conditions which will be formulated below) are also called *generalized measures* on the space X. Generalized measures have been rather extensively studied; in particular, the measures considered in the example in § 2 are employed very often in the theory of random processes. They are called measures corresponding to "white noise".

Let B be a (linear) bounded symmetric positive operator. Introduce in X the new scalar product:

$$(x, y)_- = (B\,x, y), \qquad |x|_-^2 = (B\,x, x)\,. \tag{1}$$

Generally speaking, X will be not be complete in the metric generated by this scalar product. Denote the completion of X in the norm $|\cdot|_-$ by X_-^B (this can be considered as the extension of X). X will be everywhere dense in X_-^B; X_-^B and X will coincide if B^{-1} is a bounded operator. Let X_+^B be the domain of definition of the operator $B^{-1/2}$ (it is dense in X) with the scalar product

$$(x, y)_+ = (B^{-1/2}\,x, B^{-1/2}\,y) = (B^{-1}\,x, y)\,. \tag{2}$$

The second equality in (2) requires some explanation. Note that for all $x \in X_+^B$ the scalar product (x, z), defined w.r.t. z on X can be extended by continuity in the metric of X_-^B to all of X_-^B. Indeed, let $x = B^{1/2}\,x_0$, $x_0 \in X$. Then

$$\begin{aligned}|(x, z_n - z_m)| &= |(B^{1/2}\,x_0, z_n - z_m)| = \\ &= |(x_0, B^{1/2}(z_n - z_m))| \le |x_0|\,\big(B^{1/2}(z_n - z_m), B^{1/2}(z_n - z_m)\big)^{1/2} = \\ &= |x_0| \cdot |z_n - z_m|_-\,.\end{aligned}$$

Hence, the linear functional (x, z) on X is continuous in the metric of X_-^B, so that it can be extended by continuity (in z) to X_-^B. In what follows, we will understand by (x, z), where $x \in X_+^B$ and $z \in X_-^B$, precisely this extension. The operator B can also be extended by continuity to X_-^B since

$$|B\,x|_- = \sqrt{(B\,x,\, B\,x)_-} = \sqrt{(B^2\,x,\, B\,x)} =$$
$$= \sqrt{(B^2\,B^{1/2}\,x,\, B^{1/2}\,x)} \le \sqrt{\|B^2\| \cdot (B^{1/2}\,x,\, B^{1/2}\,x)} \le \|B\| \cdot |x|_- .$$

In the sequel we will assume that B has been extended to X_-^B. In this connection we have the relations

$$B^{1/2}\,X_-^B = X\,, \qquad B^{1/2}\,X = X_+^B; \qquad B\,X_-^B = X_+^B\,.$$

The third of these follows from the first two; the second is a consequence of the definition of X_+^B. Let us prove the first. Let z be an arbitrary element in X_-^B, $z_n \in X$ and assume $|z_n - z|_- \to 0$. This means that

$$(B(z_n - z_m),\, z_n - z_m) = |B^{1/2}\,z_n - B^{1/2}\,z_m|^2 \to 0; \qquad n, m \to \infty\,.$$

But $B^{1/2}\,z_n \in X$ so that $B^{1/2}\,z$ is also in X. We return to Eq. (2). By what has been proved above, $B^{-1}\,x$ is defined for $x \in X_+^B$ and belongs to X_-^B. Hence, for $y \in X_+^B$, $(B^{-1}\,x, y)$ is also defined.

Now let a measure μ be defined on X_-^B. Then we can define in terms of this measure the characteristic functional

$$\theta_-(z) = \int e^{i(x,z)_-}\,\mu(dx)\,,$$

for $z \in X_-^B$. Since $(x, z)_- = (B\,z, x)$ and $B\,z \in X_+^B$, the measure μ on X_-^B can also be given by means of the characteristic functional

$$\theta(z) = \int e^{i(z,x)}\,\mu(dx)$$

where $z \in X_+^B$. Note that

$$\theta_-(z) = \varphi(B\,z) \quad \text{and} \quad \theta(z) = \varphi_-(B^{-1}\,z)\,.$$

From Theorem 1 §4 it follows that $\varphi_-(z)$ is the characteristic functional of a measure on X_-^B iff for each $\varepsilon > 0$ there exists a nuclear operator S on X_-^B for which Re $(\theta_-(0) - \theta_-(z)) < \varepsilon$ when $(S\,z, z)_- < 1$. We will use this result to construct an extension of X such that the given positive-definite functional $\varphi(z)$ is a characteristic functional in this extension.

Theorem 1. *Let $\theta(z)$ be a continuous positive-definite functional defined on X. Then for any nuclear operator B, $\varphi(z)$ is the characteristic functional of some measure on X_-^B.*

Proof: It follows from the continuity of $\varphi(z)$ that for any $\varepsilon > 0$ one can find a $\delta > 0$ for which Re $(\theta(0) - \theta(z)) \le \varepsilon$ when $|z|^2 < \delta$. Then for $z \in X_-$

$$\operatorname{Re}(\varphi(B(0)) - \varphi(B(z)) \le \varepsilon$$

if $(B\, z, B\, z) \leq \delta$, i.e., $\operatorname{Re}\left(\theta_-(0) - \theta_-(z)\right) \leq \varepsilon$ if $\left(\frac{1}{\delta} B\, z, z\right)_- \leq 1$. We will show that the operator $\frac{1}{\delta} B$, defined on X_-^B, is nuclear. To do this, it is sufficient to prove that B is nuclear. But

$$1)\ (B\, x, y)_- = (B^2\, x, y) = (B\, x, B\, y) = (x, B\, y)_-;$$

$$2)\ (B\, x, x)_- = (B\, x, B\, x) \geq 0\,.$$

We show finally that

$$3)\ \operatorname{tr}_- B = \sum_{k=1}^{\infty} (B\, e_k, e_k)_- < \infty\,,$$

where $\{e_k\}$ is an orthonormalized basis in X_-^B. Indeed, set $\varepsilon_k = f_k / \sqrt{\lambda_k}$, where $\{f_k\}$ is a basis taken from the eigenvectors of the operator B in X, $\lambda_k = (B\, f_k, f_k)$, $(f_k, f_k) = 1$. Then

$$\operatorname{tr}_- B = \sum_{k=1}^{\infty} (B^2\, e_k, e_k) = \sum_{k=1}^{\infty} \frac{1}{\lambda_k} (B^2\, f_k, f_k) = \sum_{k=1}^{\infty} \lambda_k = \operatorname{tr} B\,. \qquad \square$$

Remark 1. Assume the positive-definite function $\theta(z)$ satisfies the condition: for each $\varepsilon > 0$ there exists a $\delta > 0$ such that $\operatorname{Re}\left(\theta(0) - \theta(z)\right) < \varepsilon$ when $(V\, z, z) < \delta$, where V is a bounded, symmetric positive operator.

Consider the space X_-^S, where S is a symmetric positive operator commuting with V. We will find a condition which must be imposed on S in order that there exist a measure in X_-^S with the characteristic functional $\theta(z)$. Since

$$\operatorname{Re}\left(\theta_-(0) - \theta_-(z)\right) = \operatorname{Re}\left(\theta(0) - \theta(S\, z)\right)$$

for $(V\, S\, z, S\, z) = (V\, S\, z, z)_- < \delta$ and $\operatorname{tr}_- V\, S = \operatorname{tr} V\, S$, such a measure will exist if $\operatorname{tr} V\, S < \infty$. This assertion also holds when $\theta(z)$ is defined on a linear manifold dense in X and V is an unbounded operator.

Measures defined on X_-^S for some S whose characteristic functionals are defined in the scalar product of X on an everywhere dense set in X will be called *generalized measures* on X. Theorem 1 shows that a generalized measure constructed w.r.t. its characteristic functional defined in X is not unique. Let X' and X'' be two extensions of X in which the measures μ' and μ'' are defined w.r.t. the same characteristic functional $\theta(z)$. Then we can find an extension X''' which is contained in each of the extensions X' and X'' and for which $\mu'(X' - X''') = 0$, $\mu''(X'' - X''') = 0$ and μ' coincides with μ'' on X'''. This extension X''' is easy to construct: if

$$X' = X_-^{S_1} \qquad \text{and} \qquad X'' = X_-^{S_2}, \qquad \text{then } X''' = X_-^{S_1+S_2}\,.$$

The generalized measure constructed in this way is unique.

The space X_-^B allows a more convenient formulation of the condition of the Minlos-Sazonov theorem.

Remark 2. In order for the condition of Theorem 1 § 4 to hold it is necessary and sufficient that there exist a nuclear operator B such that $\theta(z)$ is continuous in the metric of X_-^B and consequently, can be extended to X_-^B. Indeed, if $\theta(z)$ is continuous in the metric of X_-^B, then for each $\varepsilon > 0$, we can find a $\delta > 0$ for which $\operatorname{Re}\left(\theta(0) - \theta(z)\right) \leq \varepsilon$ when $|z| < \delta$, i.e., $\left(\frac{1}{\delta} B z, z\right) \leq 1$ and $\frac{1}{\delta} B$ is a nuclear operator. The existence of such a B for the characteristic functional of a measure in X follows from the remark in § 4.

Chapter 2. Measurable Functions on Hilbert Space

§ 7. Measurable Linear Functionals

We consider the measurable Hilbert space $(X, \mathfrak{B})$ on which a measure μ is defined. Each continuous linear functional $\varphi(x)$ defined on X is obviously $\mathfrak{B}$-measurable. It is well known that if a sequence $\varphi_n(x)$ converges to some limit $\varphi(x)$ for all x, then this limit will also be a continuous linear functional on X. The situation changes if one requires that $\varphi_n(x)$ need not possess a limit for all x, but merely on a set D for which $\mu(D) = 1$. It is natural to call functions $\varphi(x)$, appearing as such limits, *$\mathfrak{B}$-measurable linear functionals.* As limits of sequences of measurable functions, they will also be $\mathfrak{B}$-measurable. From the relation

$$\lim_{n\to\infty} \varphi_n(\alpha\, x + \beta\, y) = \alpha \lim_{n\to\infty} \varphi_n(x) + \beta \lim_{n\to\infty} \varphi_n(y)$$

it follows that the domain of definition D of the functional $\varphi(x)$ (we assume that it is defined wherever the corresponding limit exists) is a linear manifold and that $\varphi(x)$ is a linear (additive and homogeneous) functional on D. In the sequel we will consider nondegenerate measures μ for which $\mu(L) = 0$ for any proper subspace L of X. Since $\mu(D) = 1$, D is dense in X. Hence, if $\varphi(x)$ is a functional measurable in the indicated sense, then: 1) it is defined on a $\mathfrak{B}$-measurable linear manifold D for which $\mu(D) = 1$; 2) $\varphi(x)$ is a $\mathfrak{B}$-measurable function and 3) $\varphi(x)$ is linear on D. It turns out that these conditions are also sufficient to ensure the μ-measurability of $\varphi(x)$. This follows from

Theorem 1. *If the function $\varphi(x)$ satisfies Conditions 1)—3), then there exists a sequence of continuous linear functionals $\varphi_n(x)$ for which*

$$\varphi(x) = \lim_{n\to\infty} \varphi_n(x) \qquad (\mathrm{mod}\ \mu)^1 .$$

Proof. We construct a sequence of continuous functionals $\varphi_n(x)$ which converges to $\varphi(x)$ w.r.t. μ. Since we can choose from the sequence a subsequence converging almost everywhere w.r.t. the measure μ,

---

[1] That is, this equality holds except possibly on a set of μ-measure zero. — Transl.

the theorem will then be proved. Let $S_c = \{x: |\varphi(x)| < c\} \cap D$. Since $\lim_{c\to\infty} \mu(D - S_c) = 1$, for all $\varepsilon > 0$ we can find a c and a compact $K \subset S_c$ such that $\mu(D - K) < \varepsilon$. Since S_c is convex, we can assume without loss of generality that K is a convex, symmetric set. Because $\mu(\{0\}) = 0$, it is possible to find a $\delta > 0$ for which

$$\mu(K - S_\delta(0)) > \mu(D) - \varepsilon .$$

$(S_\delta(0)$ is the sphere of radius δ and center at $0)$. For $x \in K - S_\delta(0)$ the inequality $\frac{|\varphi(x)|}{|x|} \leq \frac{c}{\delta}$ is obviously fulfilled. This inequality will also hold for $x \in L$, where L is the linear hull of the set $K - S_\delta(0)$. Thus, $\varphi(x)$ is a bounded linear functional on L. By the Hahn-Banach theorem there exists a linear extension of φ to all of X with the same modulus of continuity. Denote this extension by $\varphi_\varepsilon(x)$. Let K_1 be the convex hull of $K - S_\delta(0)$. Then

$$\mu(\{x: |\varphi_\varepsilon(x) - \varphi(x)| > 0\}) \leq \mu(X - K_1) = \mu(X) - \mu(K_1) < \varepsilon .$$

There thus exists a sequence of bounded linear functionals converging in measure μ to $\varphi(x)$. □

Remark. If a sequence of finite-dimensional subspaces L_n is such that $L_n \subset L_{n+1}$, $L_n \subset D$ and $\bigcup L_n$ is dense in D, then $\varphi(P_n x)$ converges (in μ) to $\varphi(x)$, where P_n is the projection operator onto L_n. Indeed, let K_1 be the compact constructed in the proof of the theorem. If n is chosen such that L_n generates a $\frac{\varepsilon\delta}{c}$ — net in K_1 (δ and c are also as in the proof of the theorem), then

$|\varphi(P_n x) - \varphi(x)| \leq \sup_{y\in L} \frac{|\varphi(y)|}{|y|} \cdot |P_n x - x| \leq \frac{c}{\delta} |P_n x - x| < \varepsilon$ for all $x \in K_1$. This means that for $m > n$

$$\mu(\{x: |\varphi(P_m x) - \varphi(x)| > \varepsilon\}) < \varepsilon .$$

In order to construct the space of all μ-measurable functionals it is convenient to use the characteristic functional $\theta(z)$ of the measure μ. Assume the sequence of continuous functionals (z_n, x) converges (in μ) to some measurable functional $\varphi(x)$. Then for each real t

$$\lim_{n,m\to\infty} \exp\{i\, t(z_n - z_m, x)\} = 1$$

in the sense of convergence w.r.t. the measure μ. This also implies

$$\lim_{n,m\to\infty} \theta(t(z_n - z_m)) = \lim_{n,m\to\infty} \int \exp\{i\, t(z_n - z_m, x)\}\, \mu(dx) = 1 . \quad (1)$$

Let

$$K(z) = \int (1 - \theta(t\, z)) \frac{1}{1 + |t|^2}\, dt .$$

Then a necessary and sufficient condition for the existence of the limit (w.r.t. μ) of the sequence (z_n, x) is that

$$\lim_{n,m\to\infty} K(z_n - z_m) = 0\,.$$

The necessity of this condition follows from (1) and the theorem on passage to the limit under an integral sign. To establish the sufficiency, note that

$$K(z) = \int\left[\int (1 - e^{it(z,x)}) \frac{1}{1+t^2}\,dt\right]\mu(dx) = \pi\int (1 - e^{|(z,x)|})\,\mu(dx)\,.$$

Therefore, for any $\varepsilon > 0$

$$\mu\left(\{x\colon |z_n - z_m, x| > \varepsilon\}\right) \le \frac{1}{\pi} K(z_n - z_m)\,(1 - e^{-\varepsilon})^{-1}\,,$$

whence also follows the convergence of (z_n, x) (w.r.t. μ). Since

$$\begin{aligned} K(z_1 + z_2) &= \pi \int (1 - e^{-|(z_1,x)+(z_2,x)|})\,\mu(dx) \le \\ &\le \pi\int (1 - e^{-|(z_1,x)|-|(z_2,x)|})\,\mu(dx) \le \pi\int (1 - e^{-|(z_1,x)|})\,\mu(dx) + \\ &\quad + \pi\int (1 - e^{-|(z_2,x)|})\,\mu(dx) = K(z_1) + K(z_2)\,, \end{aligned}$$

X can be considered to be a metric space with metric

$$r(x, y) = K(x - y)\,.$$

Let $\tilde{X}$ denote the completion of X in the metric r. Each element of $\tilde{X}$ can be associated with some μ-measurable functional $\varphi(x)\colon \tilde{x} \overset{S}{\leftrightarrow} \varphi$ if there exists a sequence z_n in X for which $r(z_n, \tilde{x}) \to 0$ and $(z_n, x) \to \varphi(x)$ (in μ). Denote the space of all μ-measurable linear functionals by $L(\mu)$. Functionals coinciding a.e. w.r.t. μ will be identified with each other. Then the correspondence S between X and $L(\mu)$ is one-to-one.

If the distance

$$r(\varphi_1, \varphi_2) = \pi\int (1 - e^{-|\varphi_1(x)-\varphi_2(x)|})\,\mu(dx)$$

is introduced in $L(\mu)$, then the relation quoted above is isometric. Hence, it is natural to simply identify the spaces $\tilde{X}$ and $L(\mu)$, which we will always do in the sequel.

We note another feature of the space $\tilde{X}$ with the metric r. The characteristic functional of the measure μ can be extended by continuity in the metric r to all of $\tilde{X}$. This extension can be written in the form

$$\theta(\varphi) = \int e^{i\varphi(x)}\,\mu(dx)$$

(here φ is a measurable functional considered as an element of $\tilde{X} = L(\mu)$). We shall show that in a certain sense, $\tilde{X}$ is the largest space to which $\theta(z)$ can be extended by continuity.

Let Y be a linear metric space with metric ϱ for which $\varrho(x, y) = \varrho(0, x - y)$ and $\theta(z)$ is continuous in the metric ϱ on X and extendible by continuity to Y. Since θ is continuous in the metric ϱ, for each $\varepsilon > 0$ one can find a $\delta > 0$ such that $\operatorname{Re}\left(1 - \varphi(z)\right) < \varepsilon$ when $\varrho(0, z) < \delta$. Then, using the inequality

$$|\theta(z_1) - \theta(z_2)| \leq \int |1 - e^{i(z_1 - z_2, x)}| \, \mu(dx) \leq$$
$$\leq \int \sqrt{2\left(1 - \cos(z_1 - z_2, x)\right)} \, \mu(dx) \leq \sqrt{2 \operatorname{Re}\left(1 - \theta(z_1 - z_2)\right)}$$

we find that for $\operatorname{Re}\left(1 - \theta(z)\right) < \varepsilon$

$$\operatorname{Re}\left(1 - \theta(n z)\right) \leq \sum_{k=1}^{n} |\theta((k-1) z) - \theta(k z)| \leq n \sqrt{\varepsilon}\,.$$

Hence, for $\varrho(0, z) < \delta$

$$K(z) = \int_{-\infty}^{\infty} \operatorname{Re}\left(1 - \theta(t z)\right) \frac{dt}{1 + t^2} = \int_{|t| \leq n} \operatorname{Re}\left(1 - \theta(t z)\right) \frac{dt}{1 + t^2} +$$
$$+ \int_{|t| > n} \operatorname{Re}\left(1 - \theta(t z)\right) \frac{dt}{1 + t^2} \leq \pi n \sqrt{\varepsilon} + \frac{4}{n}\,.$$

Thus, if $\varrho(z_n, z_m) \to 0$, then $K(z_n - z_m) \to 0$ so that Y can be imbedded isometrically in some subset of $\tilde{X}$.

The space $\tilde{X}$ is considerably larger than X since it contains, for example, the spaces X_{-}^{B} obtained by completing X in the scalar product $(x, y)_{-} = (B x, y)$, where B is a nuclear operator for which $\varphi(z)$ is continuous in the scalar product $(B z, z)$.

We now consider inhomogeneous, linear, μ-measurable functionals. This is a natural designation for $\mathfrak{B}$-measurable functions $\varphi(x)$ for which there exists a sequence $z_n \in X$ and a sequence of constants α_n such that

$$\varphi(x) = \lim_{n \to \infty} [\alpha_n + (z_n, x)] \qquad (\operatorname{mod} \mu)\,. \tag{2}$$

If for a given φ there exists a bounded sequence α_n such that (2) holds, then $\varphi(x) = \alpha + \varphi_0(x)$, where α is a constant and $\varphi_0(x)$ is a linear μ-measurable functional. This follows from the fact that if the sequence is such that $\alpha_{n_k} \to \alpha$, then $(z_{n_k}, x) \to \varphi(x) - \alpha$ (mod μ). It turns out also in the general case that the structure of inhomogeneous linear μ-measurable functionals is the same.

Theorem 2. *For any inhomogeneous linear μ-measurable functional $\varphi(x)$ there exists a constant α and a μ-measurable linear functional $\varphi_0(x)$ for which*

$$\varphi(x) = \alpha + \varphi_0(x)\,. \tag{3}$$

Proof. If a representation of the form (2) for $\varphi(x)$ exists with bounded α_n, then the possibility of writing (3) is already proved. Now let φ be represented by (2), where $|\alpha_n| \uparrow \infty$. Then it follows from (2) that

$$\lim_{n\to\infty}\left(\frac{z_n}{-\alpha_n}, x\right) = 1 .$$

Let N_m be such that for $n \geq N_m$

$$\mu\left(\left\{x: \left|\left(\frac{z_n}{-\alpha_n}, x\right) - 1\right| > \frac{1}{2^m}\right\}\right) < \frac{1}{2^m} .$$

Then

$$\lim_{k\to\infty} \alpha_k \left[\left(\frac{z_{n_k}}{-\alpha_{n_k}}, x\right) - 1\right] = 0 \qquad (\mathrm{mod}\,\mu) ,$$

provided that $n_k \geq N_{m_k}$, where m_k is the integral part of $|\alpha_k| + 1$. In fact, for any $\varepsilon > 0$

$$\mu\left(\bigcup_{k=l}^{\infty}\left\{x: \left|\left(\frac{z_{n_k}}{-\alpha_{n_k}}\right) - 1\right| > \frac{\varepsilon}{|\alpha_{n_k}|}\right\}\right) \leq$$

$$\leq \sum_{k=l}^{\infty} \mu\left(\left\{x: \left|\left(\frac{z_{n_k}}{-\alpha_{n_k}}\right) - 1\right| > \frac{\varepsilon}{|\alpha_{n_k}|}\right)\leq \sum_{m=l}^{\infty} 2^{-m} ,$$

if $\varepsilon/|m_l| > 2^{-m_\varepsilon}$.

Hence,

$$\lim_{k\to\infty} [\alpha_k + (z_k, x)] = \lim_{k\to\infty}\left[\alpha_k\left(\frac{z_{n_k}}{-\alpha_{n_k}}, x\right) + (z_k, x)\right] =$$

$$= \lim_{k\to\infty}\left(z_k - \frac{\alpha_k}{\alpha_{n_k}} z_{n_k}, x\right).$$

This means that in this case $\varphi(x)$ is itself a linear μ-measurable functional. □

Remark. From the proof of Theorem 2 it follows that there exist two possibilities for the measure μ: a) the constant $\alpha \neq 0$ cannot be represented as the limit w.r.t. μ of a sequence of linear functionals, b) the classes of inhomogeneous linear and linear μ-measurable functionals coincide.

§ 8. Measurable Linear Operators

As in the case of measurable functionals, measurable linear operators are naturally defined as the limit (w.r.t. μ) of a sequence of continuous operators. Since we can consider either strong or weak convergence of a sequence $A_n x$, measurability can be defined in either the strong or weak sense. Hence, the operator A will be called *strongly* (*weakly*) *measurable w.r.t.* μ if there exists a sequence of continuous linear operators A_n such that $A_n x$ converges strongly (weakly) to $A x$ (mod μ). Clearly, a strongly measurable operator is also weakly measurable. Assume A is weakly measurable. Denote by D_A the set of all x for which the weak limit of the sequence $A_n x$ exists. Then, denoting by N a denumerable dense set in X, we have

$$D_A = \left\{x: \sup_n |A_n x| < \infty;\ \lim_{n\to\infty} (z, A_n x) \text{ exists}, z \in N\right\}.$$

We see that D_A is measurable. It is also clear that D_A is a linear manifold. For all $x \in D_A$, the weak limit $A x = \lim\limits_{n\to\infty} A_n x$ exists; moreover, $A(\alpha x + \beta y) = \alpha A x + \beta A y$ for all real α and β and $x, y \in D_A$. Finally, $\mu(D_A) = 1$. We will show that the validity of the listed conditions is even sufficient to ensure that the operator A is strongly measurable; be the same token, we prove that the notions of strong and weak measurability are equivalent.

Theorem 1. *Suppose that on the measurable linear manifold D_A for which $\mu(D_A) = 1$ there is defined a measurable function $A x$ with values in X which satisfies for all $x, y \in D_A$ and real α, β the relation $A(\alpha x + \beta y) = \alpha A x + \beta A y$. Then there exists a sequence of continuous linear operators A_n for which*

$$A x = \lim_{n\to\infty} A_n x \qquad (\text{mod}\ \mu)\ .$$

Proof. We note that $|A x|$ is a measurable function. Hence,

$$\lim_{c\to\infty} \mu(\{x: |A x| > c\}) = 0$$

which implies that for all $\varepsilon > 0$ one can find a compact K for which $|A x| \leq c$ for $x \in K$ and $\mu(X - K) < \varepsilon$. This compact can be assumed to be a convex, centrally-symmetric set. Choose δ as in the proof of Theorem 1 § 7. Let K_1 and L have the same meaning as in that theorem. Then $\frac{|A x|}{|x|} \leq \frac{c}{\delta}$. Let N be a finite-dimensional subspace of L such that $N \cap K_1$ forms a $\frac{\delta}{c}\,\varepsilon$-net in K_1. We construct an operator A_N as follows: for $x \in N$, $A_N x = A x$; however, if y is orthogonal to N, then $A_N y = 0$. By such an extension of A from N to all of X we do not increase the modulus of continuity. Hence, denoting by P_N projection onto N, we have $|A_N x - A x| \leq \frac{c}{\delta} |x - P_N x| < \varepsilon$ for $x \in K_1$. Thus,

$$\mu(\{|A_N x - A x| > 2\varrho\}) < \varepsilon\,.$$

Choosing sequences $\varepsilon \to 0$ and $\varrho \to 0$, we construct a sequence of bounded linear operators which converges to A in measure and from such a sequence we can choose a subsequence converging almost everywhere. □

In what follows we will use simply the term measurable linear operator without referring to the adjectives strong or weak.

To describe the construction of the μ-measurable linear operator A we note that for all $z \in X$ the expression

$$(A\, x, z) = \varphi_z(x)$$

as function of x will be a linear μ-measurable functional. Denote by $\tilde{x}(z)$ the element from $\tilde{X}$ corresponding to this functional. It is clear that $x(z)$ depends linearly on z. Hence, with each μ-measurable linear operator A we can associate some linear operator A^* acting on X into $\tilde{X}$; these two operators are connected by

$$(A\, x, z) = [A^* z]\,(x) \qquad (\operatorname{mod} \mu)\,. \tag{1}$$

$[A^* z]$ is a μ-measurable linear functional. The operator $A^* z$ is such that $A^* z \to 0$ in the metric of $\tilde{X}$ when $z \to 0$ weakly. Moreover, for almost all x (w.r.t. μ) there exists a constant γ_x such that

$$|[A^* z_1]\,(x) - [A^* z_2]\,(x)| \leq \gamma_x\, |z_1 - z_2|\,. \tag{2}$$

If now an arbitrary linear operator A^* is given mapping X into $\tilde{X}$ and satisfying the quoted continuity condition and (2), then to this operator there corresponds a μ-measurable operator A for which (1) holds. Indeed, let D be the set of all x for which $[A^* z]\,(x)$ is continuous w.r.t. z. It follows from (2) that $\mu(D) = 1$. It's easy to see that D is a linear manifold. For all $x \in D$ there exists a vector $A\, x \in X$ such that (1) is satisfied. $A\, x$ is a homogeneous additive operator on D. The measurability of $A\, x$ follows from the equality

$$A\, x = \sum\, [A^* e_k]\,(x)\,,$$

which holds for any orthonormalized basis with $x \in D$, and the measurability of $[A^* e_k]\,(x)$.

We consider the notion of an absolutely measurable linear operator. A measurable linear operator A is called *absolutely measurable* if for every measurable linear functional $l(x)$, the expression $l(A\, x)$ is also a measurable linear functional. This idea can be understood in two senses. Firstly, since there exists a sequence z_n such that $K(l - z_n) \to 0$, we can view $l(A\, x)$ as the limit (in μ) of the sequence of measurable linear functionals $l_n(x) = (z_n, A\, x)$. Secondly, $l(A\, x)$ can be viewed as the ordinary superposition of two measurable functions. It will also be measurable. The additivity and homogeneity condition holds in the domain of definition for this function. This domain of definition is the set $\{x\colon A\, x \in D_l\}$, where D_l is the domain of definition of $l(x)$. If Δ_A

denotes the range of the operator A, then $l(A\,x)$ is a measurable functional iff

$$\mu\left(A^{-1}(\Delta_A \cap D_l)\right) = 1\,.$$

However, since D_l can be taken as any measurable linear manifold L for which $\mu(L) = 1$, it is necessary that the condition $\mu\left(A^{-1}(\Delta_A \cap L)\right) = 1$ hold when $\mu(L) = 1$. Using Theorems 1 and 2 we can convince ourselves of the equivalence of both approaches to the measurability of $l(A\,x)$. We now describe the structure of an absolutely measurable operator. Note that for any absolutely measurable operator A, the convergence w.r.t. μ of a sequence of measurable linear functionals $l_n(x)$ to $l(x)$ implies the convergence in μ of the sequence $l_n(A\,x)$ to $l(A\,x)$. To show this it is sufficient to consider almost everywhere convergence instead of convergence in measure. Then we can find a linear manifold L, $\mu(L) = 1$, for which $l_n(x) \to l(x)$ for $x \in L$. This means that $l_n(A\,x) \to l(A\,x)$ for all x for which $A\,x \in L$ and this set has measure 1 due to the absolute measurability of A. Hence, we can associate with the absolutely measurable operator A an operator A^* mapping $\tilde{X}$ into X according to $[A^*\,l]\,(x) = l(A\,x)$. From the above it follows that this operator is continuous in the metric r since convergence of functionals in this metric is equivalent to their convergence in μ. We now show that the converse is also true: if A is a measurable linear operator for which the operator A^*, defined by $A^*\,z(x) = (z, A\,x)$ for all $z \in X$, is extended by continuity in the metric r to all of $\tilde{X}$, then A is absolutely measurable. If A^* can be extended by continuity to $\tilde{X}$, then $\varphi(A^*\,z)$ will be an r-continuous positive-definite functional. Hence, the series

$$A\,x = \sum_{k=1}^{\infty} [A^*\,e_k]\,(x)\,e_k$$

converges almost everywhere w.r.t. μ for any orthonormalized basis $\{e_k\}$; moreover, $\varphi(A^*\,z)$ will be the characteristic functional of the measure $\mu(A^{-1}\,dx)$. Let l be a measurable functional, $\{f_k\}$ an orthonormalized basis in D_l. Then

$$A\,x = \sum_{k=1}^{\infty} (A\,x, f_k)\,f_k = \sum_{k=1}^{\infty} (A^*\,f_k, x)\,f_k\,.$$

Let P_n be the projection operator on the subspace spanned by $f_1, \ldots, f_n$. We will show that $l(P_n\,A\,x)$ converges (in μ) to some limit. In fact,

$$l(P_n\,A\,x) = \sum_{j=1}^{n} (A^*\,f_j, x)\,l(f_j) = \left[A^* \sum_{j=1}^{n} l(f_j)\,f_j\right](x)$$

and since $\left(\sum_{j=1}^{n} l(f_j)\, f_j,\, x\right) = l(P_n x)$ converges (in μ) to $l(x)$ on the basis of Theorem 1, we have

$$\left[A^* \sum_{j=1}^{n} l(f_j)\, f_j\right](x) \to (A^* l)\,(x)$$

in μ because of the continuity of A^* in $\tilde{X}$. Therefore,

$$l(A\,x) = [A^* l]\,(x)$$

is a measurable linear functional for any linear functional l. The absolute measurability of A is proved.

We consider measurable linear mappings of a Hilbert space X into a second Y. We will only treat mappings which are strongly measurable. It will turn out that the study of such mappings can easily be reduced to the study of measurable mappings from X into X, i.e., measurable linear operators. Indeed, let R be an isometric one-to-one mapping of Y onto X (we assume both spaces are separable). Let V be a measurable mapping of X into Y; then we can find a sequence of continuous linear mappings V_n for which $V_n\, x \to Vx$ (in Y) w.r.t. the measure μ. But $V_n\, R$ is already a sequence of continuous linear mappings of X into X which converges (in μ) to $V\, R$. This implies that $V\, R$ is a measurable linear operator. Conversely, if U is a measurable linear operator from X into X, then $U\, R^{-1}$ is a measurable linear mapping of X into Y. Hence, all measurable linear mappings from X into Y are completely characterized.

Denote by ν the measure defined in Y by means of $\nu(E) = \mu(V^{-1}(E))$, where V is a measurable linear mapping from X into Y. We will now find the characteristic functional of the measure ν. To do this we need the notion of the transformation conjugate to V. Let D be the domain of definition of the map V. $\mu(D) = 1$. The expression $(V\,x, y)$ is defined for all $y \in Y$ and $x \in D$ and is a measurable linear functional on D. Hence, there exists an element $l_y \in \tilde{X}$ such that $(V\,x, y) = l_y(x)$. Put $l_y = V^* y$. $V^* y$ defines a homogeneous and additive mapping of Y into X which is continuous in the following sense: $r(V^* y_1, V^* y_2) \to 0$ for $|y_1 - y_2| \to 0$. This mapping is said to be *conjugate* to V. Of special interest is the case where V^* can be considered as a measurable (w.r.t. ν) mapping from Y into X. Let $\{e_k\}$ be some orthonormalized basis in X. In order that $V^* y$ be a member of X it is necessary and sufficient that

$$V^* y = \sum_{k=1}^{\infty} (V^* y, e_k)\, e_k = \sum_{k=1}^{\infty} (y, V\, e_k)\, e_k$$

and the series $\sum_k (y, V e_k)^2$ converge a.e. w.r.t. the measure ν. The latter condition is equivalent to the a.e. convergence w.r.t. μ of $\sum (V x, V e_k)^2$. Finally, we shall find $\theta_\nu(y)$, the characteristic function of the measure ν. The symbol $\theta_\mu(l)$ will denote the extension of the characteristic functional $\theta_\mu(z)$ to $\tilde{X}$ by continuity. Then

$$\theta_\nu(y) = \int e^{i(y, V x)} \mu(dx) = \int e^{i(V^* y, x)} \mu(dx) = \theta_\mu(V^* y) .$$

§ 9. Measurable Polynomial Functions

We first consider polynomials with numerical (real) values. Let X^{*k} denote the linear space of continuous symmetric k-linear real functions $\xi^{(k)}(x_1, x_2, \ldots, x_k)$, defined for $x_1, \ldots, x_k \in X$, linear in each argument when the rest are fixed, invariant under permutations of the arguments and satisfying the condition

$$\lim_{x_1 \to 0, \ldots, x_k \to 0} \xi^{(k)}(x_1, \ldots, x_k) = 0 .$$

A function of the form $\eta_k(x) = \xi^{(k)}(x, \ldots, x)$, where $\xi^{(k)} \in X^{*k}$, is called a *homogeneous form in x of degree k*, and a function of the form

$$\pi_n(x) = \alpha + \sum_{k=1}^{n} \eta_k(x) \tag{1}$$

a *polynomial in x of degree n* (if $\eta_n(x)$ is not identically equal to zero), where the $\eta_k(x)$ are homogeneous forms of degree k. It's easy to see that the polynomial $\pi_n(x)$ uniquely determines all of the forms $\eta_k(x)$, and these forms determine in turn the functions $\xi^{(k)}$ in X^{*k}. In the sequel, polynomials of the form (1) will be called continuous since measurable polynomials also appear here.

Let a nondegenerate measure μ be given on $(X, \mathfrak{B})$. By $\Pi_n(\mu)$ denote the set of $\mathfrak{B}$-measurable functions $\varphi(x)$ which can be represented as limits of an almost everywhere convergent sequence of continuous polynomials:

$$\varphi(x) = \lim_{m \to \infty} \pi_n^m(x) \qquad (\operatorname{mod} \mu) , \tag{2}$$

where π_n^m are continuous polynomials for all $m = 1, 2, \ldots$. The set of functions $\Pi_n(\mu)$ will be called the *space of measurable polynomials of degree no greater than n*, functions from $\bigcup \Pi_n(\mu)$ *measurable* (w.r.t. μ) *polynomials*, and the smallest n for which $\varphi \in \Pi_n(\mu)$ the *degree* of the measurable polynomial.

It turns out that the description of the space $\Pi_n(\mu)$ is quite similar to that of the space of measurable linear functionals; more precisely, measurable polynomials of degree no greater than n can be identified with measurable linear functionals on some Hilbert space with measure; this Hilbert space depends only on n, and the measure on it is expressed in a standard way by means of the original measure. Let us look at this construction.

Let X^{0k} be the linear sub-space of those k-linear functions $\xi^{(k)}$ in X^{*k} for which

$$\operatorname{tr} \xi^{(k)} * \xi^{(k)} = \sum_{i_1,\ldots,i_k=1}^{\infty} \xi^{(k)}(e_{i_1}, \ldots, e_{i_k})\, \xi^{(k)}(e_{i_1}, \ldots, e_{i_k}) < \infty . \tag{3}$$

for any orthonormalized basis $\{e_j\}$ (the sum does not in fact depend on the basis). In X^{0k} one can introduce the scalar product

$$(\xi_1^{(k)}, \xi_2^{(k)}) = \operatorname{tr} \xi_1^{(k)} * \xi_2^{(k)} = \sum_{i_1,\ldots,i_k=1}^{\infty} \xi_1^{(k)}(e_1, \ldots, e_k)\, \xi^{(k)}(e_1, \ldots, e_k) \tag{4}$$

(this scalar product also does not depend on the choice of the basis in X). We further denote by $X^{[n]}$ the direct sum of the Hilbert spaces $X^{01} + \cdots + X^{0n}$, setting the scalar product of the elements $\xi_1^{(1)} + \cdots + \xi_1^{(n)}$ and $\xi_2^{(1)} + \cdots + \xi_2^{(n)}$ equal to

$$(\xi_1^{(1)} + \cdots + \xi_1^{(n)}, \xi_2^{(1)} + \cdots \xi_2^{(n)}) = \sum_1^n (\xi_1^{(k)}, \xi_2^{(k)}) . \tag{5}$$

With the scalar product (5), $X^{[n]}$ is a Hilbert space. We now construct on $X^{[n]}$ a measure related to the measure μ. Let T_k be a mapping of X into X^{0k}: $T_k(x) = \xi_x^{(k)}$, where $\xi_x^{(k)}(x_1, \ldots, x_k) = \prod_{j=1}^{k} (x, x_j)$. This mapping is continuous since

$$\operatorname{tr} (T_k(y) - T_k(z)) * (T_k(y) - T_k(z)) = \sum_{i_1,\ldots,i_k} \left[\prod_{j=1}^{k} (y, e_{i_j}) - \prod_{j=1}^{k} (z, e_{i_j}) \right]^2 =$$

$$= \sum_{i_1,\ldots,i_k} \left[\sum_{l=1}^{k} \prod_{j=1}^{l-1} (y, e_{i_j})\, (y - z, e_{i_l}) \prod_{j=l+1}^{k} (z, e_{i_j}) \right]^2 \le$$

$$\le k^2 |y - z|^2 \max_{l \le k} \{ |y|^{2l-2} + |z|^{2l-2} \} .$$

Consequently, it is measurable: the inverse image of any Borel set in X^{0k} under the mapping $T_k(x)$ will belong to $\mathfrak{B}_X$. Assume in addition that $T_n(x)$ is a mapping of X into $X^{[n]}$ defined by the equation

$$T^{[n]}(x) = T_1(x) + \cdots + T_n(x) .$$

This mapping will also be continuous and measurable. Denote by $\mu^{[n]}$ the measure in $X^{[n]}$ into which μ is transferred by the mapping $T^{[n]}$, i.e., for any set $A^{[n]}$ from $\mathfrak{B}^{[n]}$, the σ-algebra of Borel sets in $X^{[n]}$,

$$\mu^{[n]}(A^{[n]}) = \mu(T^{[n]^{-1}}(A^{[n]})) , \tag{6}$$

where $T^{[n]^{-1}}(A^{[n]})$ is the complete inverse image of $A^{[n]}$ under the mapping $T^{[n]}$. The measure $\mu^{[n]}$ is thus completely determined by the measure μ; moreover, to $\mu^{[n]}$ one can apply all the results of the earlier theory since $X^{[n]}$ is also a separable Hilbert space.

We turn now to the continuous polynomial $\pi_n(x)$ of the form (1) for which $\alpha = 0$ and

$$\eta_k(x) = \xi^{(k)}(x, \ldots, x),$$

where $\xi^{(k)} \in X^{0k}$. The set of such polynomials will be denoted by Π_n^0. It is easy to see that

$$\eta_k(x) = \xi^{(k)}(x, \ldots, x) = \operatorname{tr} T_k(x)\, \xi^{(k)},$$

because

$$\operatorname{tr} T_k(x)\, \xi^{(k)} = \sum_{i_1,\ldots,i_k} \left[\prod_{j=1}^{k} (x, e_{i_j})\right] \xi^{(k)}(e_{i_1}, \ldots, e_{i_j}) = \xi^{(k)}(x, \ldots, x).$$

Thus,

$$\pi_n(x) = \left(T^{[n]}(x), \xi^{(1)} + \cdots + \xi^{(n)}\right).$$

Let $\xi^{[n]}$ be a point in the space $X^{[n]}$ of the form $\xi^{(1)} + \cdots + \xi^{(n)}$ and let $x^{[n]}$ be an arbitrary point of $X^{[n]}$. Then $(x^{[n]}, \xi^{[n]})$ as function of $x^{[n]}$ defines on the set $Y^{[n]} \subset X^{[n]}$ ($Y^{[n]}$ is the image of X under $T^{[n]}$) a polynomial

$$(x^{[n]}, \xi^{[n]}) = \pi_n\left(T^{[n]^{-1}}(x^{[n]})\right). \tag{7}$$

Since the measure $\mu^{[n]}$ is concentrated on $Y^{[n]}$, where $T^{[n]^{-1}}$ is defined, (7) makes sense a.e. (w.r.t. $\mu^{[n]}$) on $X^{[n]}$. Hence, one can establish a one-to-one relationship between linear functionals on $X^{[n]}$ and polynomials from $\Pi_0^{[n]}$ onto X; moreover, the convergence of linear functionals in $\mu^{[n]}$ is equivalent to that of polynomials from $\Pi_0^{[n]}$ in μ.

In the investigation of measurable linear functionals we used the characteristic functional of the corresponding measure. We now show how the characteristic functional $\theta^{[n]}(\xi^{[n]})$ of the measure $\mu^{[n]}$ can be defined. By definition

$$\theta^{[n]}(\xi^{[n]}) = \int_{X^{[n]}} \exp\{i(x^{[n]}, \xi^{[n]})\}\, \mu(dx^{[n]}) =$$

$$= \int_X \exp\{i(T^{[n]}(x), \xi^{[n]})\}\, \mu(dx) = \int \exp\left\{i \sum_1^n \xi^{(k)}(x)\right\} \mu(dx).$$

Denote by $\Pi_n'(\mu)$ the measurable polynomials from $\Pi_n(\mu)$ which are limits in measure μ of polynomials in Π_0^n. If $\varphi(x) \in \Pi_n'(\mu)$, then there exists a sequence of polynomials $\{\pi_n^k, k = 1, 2, \ldots\}$ in Π_0^n such that

$$\varphi(x) = \lim_{k\to\infty} \pi_n^{(k)}(x) \quad (\operatorname{mod} \mu). \tag{8}$$

If now $\xi_k^{[n]}$ is an element of $X^{[n]}$ for which

$$\pi_n^{(k)}(x) = \left(T^{[n]}(x), \xi_k^{[n]}\right),$$

then it follows from (8) that the limit

$$\lim_{k\to\infty} (x^{[n]}, \xi_k^{[n]})$$

exists a.e. (mod $\mu^{[n]}$). This limit, in agreement with the terminology of § 7, is a $\mu^{[n]}$-measurable linear functional on $X^{[n]}$. Assume, on the other hand, that $\xi_k^{[n]}$ is a sequence of elements from $X^{[n]}$ such that

$$\lim_{k\to\infty} (x^{[n]}, \xi_k^{[n]}) = \varphi^{[n]}(x^{[n]}) \qquad (\mathrm{mod}\, \mu)$$

exists. Then

$$\lim_{k\to\infty} (T^{[n]}(x), \xi_k^{[n]}) = \varphi(x) \qquad (\mathrm{mod}\, \mu)$$

also exists. For all k

$$\pi_n^{[k]} = (T^{[n]}(x), \xi_k^{[n]})$$

is a polynomial in Π_0^n. Hence, $\varphi \in \Pi_k'[\mu]$. Denote by $L^{[n]}(\mu^{[n]})$ the set of linear $\mu^{[n]}$-measurable functionals on $X^{[n]}$. Then there exists a one-to-one relationship $\tilde{T}^{[n]}$ between $L^{[n]}(\mu^{[n]})$ and $\Pi_n'[\mu]$ defined as follows: for $\varphi \in \Pi_n'[\mu]$ and $\varphi^{[n]} \in L^{[n]}(\mu^{[n]})$,

$$\varphi^{[n]} \to \varphi$$

if

$$\varphi^{[n]}(T^{[n]}(x)) = \varphi(x) . \tag{9}$$

This relation is linear.

Thus for the construction of $\Pi_n'[\mu]$ it is first necessary to complete $X^{[n]}$ in the metric $\varrho^{[n]}$, defined by

$$\varrho^{[n]}(\xi_1^{[n]}, \xi_2^{[n]}) = \int [1 - \theta^{[n]}(\tau(\xi_1^{[n]} - \xi_2^{[n]}))] \frac{d\tau}{1+\tau^2}, \tag{10}$$

obtaining thereby $L^{[n]}(\mu^{[n]})$, and then for each $\varphi^{[n]} \in L^{[n]}(\mu^{[n]})$ to define $\varphi(x)$ by means of (9).

Now we study how $\Pi_n'[\mu]$ and $\Pi_n[\mu]$ are related.

Theorem 1. *For any function $\varphi(x) \in \Pi_n[\mu]$ there exists a $\varphi_0(x) \in \Pi_n'[\mu]$ and a constant α such that*

$$\varphi(x) = \alpha + \varphi_0(x) .$$

Proof. We first show that for $\varphi \in \Pi_n[\mu]$ there exists a $\pi_n^{(k)} \in \Pi_0^n$ and constants α_k for which

$$\varphi(x) = \lim_{k\to\infty} [\alpha_k + \pi_n^{(k)}(x)] \qquad (\mathrm{mod}\, \mu) . \tag{11}$$

To this end it suffices to show that for any sequence of homogeneous polynomials $\eta_n^{(k)}(x)$ $(k = 1, 2, \ldots)$ of degree n we can find another sequence $\bar{\eta}_n^{(k)}(x)$ belonging to Π_0^n for which

$$\lim_{k\to\infty} [\eta_n^{(k)}(x) - \bar{\eta}_n^{(k)}(x)] = 0 \qquad (\text{mod}\, \mu)\,, \tag{12}$$

since we can then replace the homogeneous polynomials $\eta_n^{(k)}$ by $\bar{\eta}_n^{(k)}$ in the sequence $\alpha + \sum_{l=1}^{n} \eta_l^{(k)}(x)$ which converges (in μ) to $\varphi(x)$. However, Eq. (12) follows from the fact that for each k we can find a finite-dimensional subspace L_k for which

$$\mu\big(\{x\colon |\eta_n^{(k)}(x) - \eta_n^{(k)}(P_{L_k}(x))| > 2^{-k}\}\big) \leq 2^{-k}\,,$$

so that we can take

$$\eta_n^{(k)}\big(P_{L_k}(x)\big)$$

as $\bar{\eta}_n^{(k)}$. Assume now that $\pi_n^{(k)} \in \Pi_n'$ also satisfies (11). Denote by $\xi_k^{[n]}$ the element of $X^{[n]}$ for which

$$\pi_n^{(k)}(x) = \big(T^{[n]}(x), \xi_k^{[n]}\big)\,.$$

Then it follows from (11) that a.e. (mod $\mu^{[n]}$) the limit

$$\lim_{k\to\infty} [(x^{[n]}, \xi_k^{[n]}) + \alpha_k] = \varphi\big(T^{[n]^{-1}}(x^{[n]})\big) \tag{13}$$

exists. Hence, the left side of (13) is an inhomogeneous linear $\mu^{[n]}$-measurable functional on $X^{[n]}$. Theorem 2 § 7 says that this limit can be represented as the sum of a constant α and a linear $\mu^{[n]}$-measurable functional on $X^{[n]}$. Hence, the theorem is proved. □

We now turn to measurable polynomial transformations of the Hilbert space X into another Hilbert space Y. As in the case of linear mappings, considered in § 8, it suffices to study measurable polynomial mappings of X into itself.

Polynomial transformations will be those transformations $U(x)$ for which $\big(U(x), z\big)$ is a polynomial in x for each z. The degree of this function is uniformly bounded w.r.t. z, otherwise the expression

$$\big(U(x), \textstyle\sum \lambda_k z_k\big)\,,$$

where z_k is chosen so that $\big(U(x), z_k\big)$ has degree k and λ_k so that $\sum \lambda_k z_k$ converges, would not be a polynomial relative to x. Let $\big(U(x), z\big) = \pi_n(x, z)$ be a polynomial of degree no greater than n. Such a transformation will be called a *polynomial transformation of degree no greater than n.*

Then

$$\pi_n(x, z) = \big(T^{[n]}(x), \xi^{[n]}(z)\big) + \alpha(z)\,.$$

where $\xi^{[n]}(z) \in X^{[n]}$ also depends linearly on z. $\alpha(z)$ is a linear functional. Hence, the expression $(\xi^{[n]}(z), x^{[n]})$ is a linear functional relative to z and can be represented as

$$\left(\xi^{[n]}(z), x^{[n]}\right) = (z, V x^{[n]}) ,$$

where V is some linear operator acting on $X^{[n]}$ into X. This operator naturally depends on the transformation $U(x)$. Thus,

$$\left(U(x), z\right) = \pi_n(x, z) = \left(T^{[n]}(x), \xi^{[n]}(z)\right) + \alpha(z) = \left(z, V\left(T^{[n]}(x)\right)\right) + \alpha(z) .$$

$\alpha(z)$ is obviously equal to $\left(U(0), z\right)$.

We now show that for any polynomial transformation $U(x)$ of the space X into itself one can find a continuous transformation $V x^{[n]}$ of $X^{[n]}$ into X such that

$$U(x) = V\, T^{[n]}(x) + U(0) .$$

Now let $U_k(x)$ be a sequence of polynomial transformations of degree no greater than n. The mapping

$$U_0(x) = \lim_{k\to\infty} U_k(x) \qquad (\operatorname{mod} \mu) , \tag{14}$$

provided this limit (in the weak sense) exists, will be called a *μ-measurable polynomial mapping of X into X of degree $\leq n$.* Assume first that $U_k(0) = 0$. Let V_k be linear maps from $X^{[n]}$ into X for which

$$U_k(x) = V_k\, T^{[n]}(x) .$$

Then for $\mu^{[n]}$-almost all $x^{[n]}$, $\lim_{k\to\infty} V_k x^{[n]}$ exists. If we denote this limit by $V_0 x^{[n]}$, then V_0 is a $\mu^{[n]}$-measurable linear map of $X^{[n]}$ into X.

Let R_n be an arbitrary isometric map of $X^{[n]}$ onto X. The existence of such a map follows from the fact that both $X^{[n]}$ and X are separable Hilbert spaces. Then V can be represented as $R\, V_0$, where V_0 is a linear $\mu^{[n]}$-measurable transformation of $X^{[n]}$ into X. Such maps (operators) are described in § 8. Hence, we finally find that for a μ-measurable polynomial transformation $U(x)$ of degree $\leq n$ which is representable as in (14) with $U_k(0) = 0$, there exists a μ-measurable linear operator V_0 on $X^{[n]}$ such that

$$U(x) = R\, V_0\, T^{[n]}(x) . \tag{15}$$

Now assume that $U(x)$ can be represented as

$$U(x) = \lim_{k\to\infty} [U_k(x) + a_k] \qquad (\operatorname{mod} \mu) , \tag{16}$$

where $U_k(0) = 0$ and $U_k(x)$ is a polynomial map of degree $\leq n$ and $a_k \in X$. If $|a_k| \leq \gamma$, then the sequence a_k is weakly compact. Thus, proceeding if necessary to a subsequence, we can write

$$U(x) = \lim_{k\to\infty} U_k(x) + a$$

(limit in the weak sense). But then $U(x) - a$ can be represented as on the right side of (15). Assume that $|a_k| \uparrow \infty$ for $k \to \infty$. Then

$$\frac{1}{|a_k|} U_k(x) + \frac{a_k}{|a_k|} \to 0 \qquad (\text{mod}\,\mu) \text{ in the strong sense.}$$

Let M_k be the linear subspace generated by the vectors a_l for $l \geq k$. Obviously, $M_n \supset M_{n+1}$. Let $\bigcap_n M_n = M$, N_n be the orthogonal complement of M_n and N the closure of $\bigcup_n N_n$, the orthogonal complement of M. From (16) we have that for all $z \in \bigcup_n N_n$

$$\big(U(x), z\big) = \lim_{k\to\infty} \big(U_k(x), z\big) \qquad (\text{mod}\,\mu)\,. \tag{17}$$

Assume that $a_k = a_k' + a_k''$, where $a_k' \in N$, $a_k'' \in M$, $U_k'(x) \in N$, $U_k''(x) \in M$, $U_k(x) = U_k'(x) + U_k''(x)$, $U(x) = U'(x) + U''(x)$, $U'(x) \in N$ and $U''(x) \in N$. From (17) it follows that

$$U'(x) = \lim_{k\to\infty} U_k'(x) \qquad (\text{mod}\,\mu)\,. \tag{18}$$

Further, for all k one can choose numbers $n_k^{(1)}, \ldots, n_k^{(r)}$ so large that

$$\left| a_k'' - \sum_{i=1}^{r} \lambda_i \frac{a_{n_k^{(i)}}}{|a_{n_k^{(i)}}|} \right| < \varepsilon\,,$$

where ε is arbitrarily small. Thus, $U_k''(x) + a_k''$ can be approximated arbitrarily closely (in μ) by the expression

$$U_k''(x) - \sum_{i=1}^{r} \lambda_i \frac{1}{|a_{n_k^{(i)}}|} U_{n_k^{(i)}}(x)$$

which means that

$$U''(x) = \lim_{k\to\infty} \tilde{U}_k''(x) \qquad (\text{mod}\,\mu)\,,$$

where $\tilde{U}_k''(0) = 0$. Hence, in this case we can find polynomial operators $\tilde{U}_k(x)$ of degree $\leq n$ for which $\tilde{U}_k(0) = 0$ and

$$U(x) = \lim_{k\to\infty} \tilde{U}_k(x)\,.$$

Using (15) we can formulate

Theorem 2. *For any μ-measurable polynomial transformation $U(x)$ of degree $\leq n$ there exists an $a \in X$ and a $\mu^{[n]}$-measurable linear operator V^0 on $X^{[n]}$ such that*

$$U(x) = a + R\, V_0\, T^{[n]}(x)\,,$$

where R is some fixed isometric map of $X^{[n]}$ into X.

§ 10. Square-integrable Polynomials

Denote by $L_2[\mu]$ the linear space of $\mathfrak{B}$-measurable real functions $\varphi(x)$ which are square-integrable w.r.t. μ. In this section we will be interested in measures μ for which the function $(z, x)^n$ is square-integrable. The class of such measures will be denoted by M_n. Let, in addition, $M_\infty = \bigcap\limits_n M_n$. For measures $\mu \in M_\infty$ all moment functions are defined. An example of a measure belonging to M_∞ is the Gaussian measure. For measures from M_∞ we will consider the set $\Pi_n^2\,[\mu]$ of square-integrable μ-measurable polynomials of degree $\leq n$. We will also introduce the subset $\tilde{\Pi}_n^2[\mu]$ of $\Pi_n^2(\mu)$ containing those functions $\varphi(x)$ for which there exists a sequence of continuous polynomials $\pi_k(x) \in \Pi_0^n$ with

$$\lim_{k\to\infty} \int [\pi_k(x) - \varphi(x)]^2\, \mu(dx) = 0$$

(the polynomials π_k belong to Π_0^n in virtue of the fact that they and only they will belong to $L_2[\mu]$ for arbitrary $\mu \in M_n$).

We note that $\tilde{\Pi}_n^2[\mu]$ is a closed linear subspace of the Hilbert space $L_2[\mu]$ (the scalar product in $L_2[\mu]$ is introduced in the usual way:

$$(\varphi_1, \varphi_2) = \int \varphi_1(x)\, \varphi_2(x)\, \mu(dx))\ .$$

This subspace arises as the completion of the space Π_0^n in the scalar product in $L_2[\mu]$. In the same way, $\Pi_n^2[\mu]$ is also a linear subspace of $L_2[\mu]$. However, the structure of this subspace is more complicated: in order to construct it, it is first necessary to construct $\Pi_n[\mu]$, the linear space of μ-measurable polynomials of degree $\leq n$ and then to find the intersection $\Pi_n[\mu] \cap L_2[\mu]$. To construct $\tilde{\Pi}_n^2[\mu]$ it suffices to know the moment functions of the measure μ up to and including $2n$-th order since the scalar product on Π_0^n is expressed by means of these functions. In fact, let $\sigma_k\,(z_1, \ldots, z_k)$ be the k-th moment function of μ (as defined in § 3). Moreover, let $\hat{\pi} \in \Pi_0^n$,

$$\pi(x) = \alpha + \sum_{k=1}^{n} \eta_k(x) \quad \text{and} \quad \hat{\pi}(x) = \hat{\alpha} + \sum_{k=1}^{n} \hat{\eta}_k(x)\,,$$

where $\eta_k(x) = \xi^{(k)}(x, \ldots, x)$, $\hat{\eta}_k(x) = \hat{\xi}^{(k)}(x, \ldots, x)$ and $\xi^{(k)}$ and $\hat{\xi}^{(k)}$ are homogeneous forms in $X^{\circ k}$. Then

$$\int \pi(x)\, \hat{\pi}(x)\, \mu(dx) = \alpha\hat{\alpha} + \hat{\alpha} \int \sum_1^n \xi^{(k)}(x, \ldots, x)\, \mu(dx) +$$

$$+ \alpha \int \sum_1^n \hat{\xi}^{(k)}(x, \ldots, x)\, \mu(dx) + \sum_{k,j=1}^n \int \xi^{(k)}(x, \ldots, x)\, \hat{\xi}^{(j)}(x, \ldots, x)\, \mu(dx)\,.$$

To calculate these integrals we note that for $\xi^{(l)} \in X^{0l}$ and an arbitrary orthonormalized basis $\{e_k\}$ in X

$$\int \xi^{(l)}(x, \ldots, x)\, \mu(dx) = \int \sum_{i_1, \ldots, i_l} \xi^{(l)}(e_{i_1}, \ldots, e_{i_l})\, (x, e_{i_1}) \cdots (x, e_{i_l})\, \mu(dx) =$$

$$= \sum_{i_1, \ldots, i_l} \xi^{(l)}(e_{i_1}, \ldots, e_{i_l})\, \sigma_l(e_{i_1}, \ldots, e_{i_l}) = \operatorname{tr} \xi^{(l)} * \sigma_l;$$

this formula is correct whenever there exists a nonnegative form $\hat{\xi}^{(m)} \in X^{\circ m}$ for which $|\xi^{(l)}(x, \ldots, x)| \le \hat{\xi}^{(m)}(x, \ldots, x)$ and

$$\int \hat{\xi}^{(m)}(x, \ldots, x)\; \mu(dx) < \infty;$$

the series in the formula converges and its sum does not depend on the choice of the basis.

If $\xi^{(k)} \times \hat{\xi}^{(j)}$ denotes a $(k + j)$-linear form defined by

$$\xi^{(k)} \times \hat{\xi}^{(j)}(x_1, \ldots, x_{k+j}) = \xi^{(k)}(x_1, \ldots, x_k) \cdot \hat{\xi}^{(j)}(x_{k+1}, \ldots, x_{k+j})\,,$$

ξ_0 is a constant, $\sigma_0 = 1$ and $\operatorname{tr} \xi^{(0)} * \sigma_0 = \xi^{(0)}$, then setting $\xi_0 = \alpha$ and $\hat{\xi}_0 = \hat{\alpha}$ we have

$$\int \pi(x)\, \hat{\pi}(x)\, \mu(dx) = \sum_{k,j=0}^n \operatorname{tr}\, [\xi^{(k)} \times \hat{\xi}^{(j)} * \sigma_{k+j}]\,. \tag{1}$$

This formula expresses the scalar product in Π_0^n by means of moment functions.

It is convenient to associate with polynomials from Π_0^n a pair $(\xi^{(0)}; \xi^{[n]})$, which corresponds to the polynomial $\pi(x) = \xi^{(0)} + \sum_1^n \xi^{(k)}(x, \ldots, x)$ if $\xi^{[n]} = \{\xi^{(1)}, \ldots, \xi^{(n)}\}$. The Hilbert space of such pairs will be denoted by $X_0^{[n]}$; we define the scalar product of the pairs $(\xi_1^{(0)}; \xi_1^{[n]})$ and $(\xi_2^{(0)}; \xi_2^{[n]})$ as

$$\xi_1^{(0)}\, \xi_2^{(0)} + (\xi_1^{[n]}, \xi_2^{[n]}),\ \xi_1^{[n]}, \xi_2^{[n]} \in X^{[n]};$$

for $\xi_1^{[n]}$ and $\xi_2^{[n]}$ the scalar product is that in $X^{[n]}$. The left side of (1) can be viewed as a bilinear form in the space $X_0^{[n]}$. We will write it as $B_0^{[n]}$:

$$B_0^{[n]}((\xi_1^{(0)}; \xi_1^{[n]}), (\xi_2^{(0)}; \xi_2^{[n]})) = \sum_{k,j=0}^n \operatorname{tr}\, [\xi_1^{(k)} \times \xi_2^{(j)} * \sigma_{k+j}]\,, \tag{2}$$

if $\xi_1^{[n]} = \{\xi_1^{(1)}, \ldots, \xi_1^{(n)}\}$, $\xi_2^{[n]} = \{\xi_2^{(1)}, \ldots, \xi_2^{(n)}\}$.

Hence, the construction of the space $\tilde{\Pi}_n^2[\mu]$ leads to the completion of the space $X_0^{[n]}$ in the scalar product generated by the bilinear form $B_0^{[n]}$ (the nonnegativity of this form follows from (1); if we identify polynomials which are equal w.r.t the measure μ, then this form will even be positive). The operation of completion of a linear set in the scalar product generated by a bilinear form can be carried out rather easily so that the problem of the construction of $\tilde{\Pi}_n^2[\mu]$ causes in principle no difficulty, provided the moment functions are known. To construct $\Pi_n^2[\mu]$, however, we must first construct $\Pi_n[\mu]$; this space is to be constructed as the completion of Π_0^n in a rather complicated metric, whose expression — even if the characteristic functional of the measure is known — cannot be given constructively.

Unfortunately, simple examples show that $\Pi_n^2[\mu]$ and $\tilde{\Pi}_n^2[\mu]$ cannot coincide.

Example. Let $\lambda_k(\tau)$ be a sequence of functions on $(-\infty, \infty)$ for which the following conditions are satisfied: a) $\lambda_k(\tau) \geq 0$, b) $\int \lambda_k(\tau)\, d\tau = 1$, c) $\int \tau \lambda_k(\tau)\, d\tau = 0$ and d) $\int\limits_{|\tau-1|>\frac{1}{k}} \lambda_k(\tau)\, d\tau \leq 1/k^2$; $\int \tau^2 \lambda_k(\tau)\, d\tau = \gamma_k < \infty$.

We introduce an arbitrary orthonormalized basis $\{e_k\}$ in X and put

$$\theta(z) = \prod_{k=1}^{\infty} \int e^{i\beta_k \tau(z, e_k)} \lambda_k(\tau)\, d\tau\,, \tag{3}$$

where the $\beta_k \geq 0$ are chosen so that $\sum \beta_k^2 \gamma_k < \infty$. To show that this product converges, we note that

$$\left|1 - \int e^{i\beta_k \tau(z,e_k)} \lambda_k(\tau)\, d\tau\right| = \left|\int \left(1 - e^{i\beta_k\tau(z,e_k)} + i\beta_k\tau(z, e_k)\right) \lambda_k(\tau)\, d\tau\right| \leq$$
$$\leq \frac{1}{2} \int \beta_k^2 \tau^2(z, e_k)^2 \lambda_k(\tau)\, d\tau = \frac{1}{2} \beta_k^2 \gamma_k(z, e_k)^2\,,$$

so that

$$\sum_{k=1}^{\infty} \left|1 - \int e^{i\beta_k\tau(z,e_k)} \lambda_k(\tau)\, d\tau\right| \leq \frac{1}{2} \sum \beta_k^2 \gamma_k(z, e_k)^2 < \infty\,.$$

From the convergence of the last series follows that of the product (3). Moreover,

$$\operatorname{Re}\left(1 - \theta(z)\right) \leq |1 - \theta(z)| \leq \sum_{k=1}^{\infty} \left|1 - \int e^{i\tau\beta_k(z,e_k)} \lambda_k(\tau)\, d\tau\right| \leq$$
$$\leq \frac{1}{2} \sum_{k=1}^{\infty} \beta_k^2 \gamma_k(z, e_k)^2 = (S z, z)\,.$$

On the right side of this equation stands a nuclear operator:

$$\sum_{j=1}^{\infty} \sum_{k=1}^{\infty} \beta_k^2 \gamma_k (e_j, e_k)^2 = \sum_{k=1}^{\infty} \beta_k^2 \gamma_k \sum_{j=1}^{\infty} (e_k, e_j)^2 = \sum_{k=1}^{\infty} \beta_k^2 \gamma_k < \infty .$$

Hence, for $\theta(z)$ the conditions of the Minlos-Sazonov theorem are satisfied (the positive-definiteness of $\theta(z)$ follows from the fact that it is the limit of a product of positive-definite functions). Let μ be a measure with characteristic functional $\theta(z)$. For μ the first two moment functions are defined:

$$\sigma_1(z) = \sum_{k=1}^{\infty} \int \beta_k \tau (z, e_k) \lambda_k(\tau) \, d\tau = 0 ,$$

$$\sigma_2(z_1, z_2) = \sum_{k \neq j} \beta_k \beta_j \int \int (z, e_k) (z, e_j) \tau_1 \tau_2 \lambda_k(\tau_1) \lambda_j(\tau_2) \, d\tau_1 \, d\tau_2 +$$

$$+ \sum_{k=1}^{\infty} \int (z, e_k)^2 \beta_k^2 \tau^2 \lambda_k(\tau) \, d\tau =$$

$$= \sum_{k=1}^{\infty} \beta_k^2 \gamma_k (z, e_k)^2 .$$

Then the space of square-integrable μ-measurable functionals will contain the functional

$$\varphi_0(x) = \lim_{k \to \infty} \left(x, \frac{1}{\beta_k} e_k \right) = 1 \qquad (\mathrm{mod}\, \mu) .$$

However, this functional is orthogonal in $L_2[\mu]$ to all linear functionals (z, x) which implies that it is not contained in the closure of the set of linear functionals in $L_2[\mu]$.

Now assume that the functions $\lambda_k(\tau)$ are such that $\lambda_k(\tau) \lambda_k(-\tau) = 0$. Let $\bar{\mu}$ be a measure with characteristic functional $\theta(-z)$. From the condition imposed on $\lambda_k(\tau)$ it is easy to see that there exist in X two closed sets F and $\bar{F}$ which are symmetric relative to $0 \in X$, have no common points and for which $\mu(F) = 1$ and $\bar{\mu}(\bar{F}) = 1$, i.e., μ and $\bar{\mu}$ are mutually symmetric measures. Let

$$\tilde{\mu} = \frac{1}{2} (\mu + \bar{\mu}) .$$

Then

$$\tilde{\varphi}_0(x) = \lim_{k \to \infty} \left(x, \frac{1}{\beta_k} e_k \right) = \begin{cases} 1: & x \in F \\ -1; & x \in \bar{F} \end{cases} \qquad (\mathrm{mod}\, \mu) .$$

It is clear that

$$\int (z, x) \tilde{\varphi}_0(x) \mu(dx) = 0 \quad \text{for all } z \in X .$$

and

$$\int \tilde{\varphi}_0(x)\, \tilde{\mu}(dx) = \frac{1}{2} - \frac{1}{2} = 0 \, .$$

Hence, $\tilde{\varphi}_0(x)$ is orthogonal to each polynomial in Π_0^1 (w.r.t. $\tilde{\mu}$) which implies that $\varphi_0 \notin \tilde{\Pi}_1^2[\mu]$; nevertheless $\tilde{\varphi}_0 \in \Pi_1^2[\mu]$.

At the present time, conditions ensuring the equality of $\Pi_n^2[\mu]$ and $\tilde{\Pi}_n^2[\mu]$ are unknown. However, for measures from M_∞, one can find a condition under which $\bigcup_n \tilde{\Pi}_n^2[\mu]$ is dense in $L_2[\mu]$.

We remark that when this condition is satisfied it is not necessary to consider the totality of all $\mathfrak{B}$-measurable functions for the construction of $L_2[\mu]$ but merely $\bigcup \Pi_0^n$ and then this linear space of polynomials is closed in the scalar product of $L_2[\mu]$.

Theorem 1. *Assume that for each $z \in X$ we can find a $\delta(z)$ such that $\theta(\zeta z)$, where $\theta(z)$ is the characteristic functional of the measure μ, can be analytically continued onto the complex plane for $|\zeta| < \delta(z)$. Then the set of all continuous polynomial functions is everywhere dense in $L_2[\mu]$.*

Proof. Assume that $\eta(x) \in L_2[\mu]$ and that for all z and $n \geq 0$

$$\int \eta(x)\, (z, x)^n\, \mu(dx) = 0 \, .$$

Then $(z, x)^n \in L_2[\mu]$ follows from the equality

$$\int (z, x)^{2n}\, \mu(dx) = (-1)^n \frac{d^{2n}}{d\zeta^{2n}}\, \theta(\zeta z)\Big|_{\zeta=0} \, .$$

Choose $\varepsilon > 0$ so that $2\varepsilon < \delta(z)$. Then

$$\int e^{2\varepsilon(z,x)} \mu(dx) < \infty \qquad \text{and} \qquad \int e^{-2\varepsilon(z,x)}\, \mu(dx) < \infty$$

since these integrals are equal to the corresponding value of the extension of $\theta(\zeta z)$ at the points $-2i\varepsilon$ and $2i\varepsilon$. Hence,

$$\int \eta(x)\, e^{i\varepsilon(z,x)}\, \mu(dx) = \lim_{n\to\infty} \int \eta(x) \sum_{k=0}^{n} \frac{(i\,\varepsilon(z, x))^k}{k!}\, \mu(dx) \, .$$

A limit passage under the integral sign is possible since the integrand has an integrable majorant:

$$|\eta(x)|\, e^{\varepsilon|(z,x)|} \leq \frac{1}{2} \left[|\eta(x)|^2 + e^{2\varepsilon(z,x)} + e^{-2\varepsilon(z,x)}\right] .$$

Differentiating the relation

$$\int \eta(x)\, e^{i\varepsilon(z,x)}\, \mu(dx) = 0$$

n times w.r.t. ε, we obtain

$$\int \eta(x)\, e^{i\varepsilon(z,x)} (z, x)^n\, \mu(dx) = 0 \, .$$

Using the same reasoning as above, we convince ourselves that

$$\int \eta(x)\, e^{i\varepsilon(z,x)}\, e^{i\varepsilon(z,x)}\, \mu(dx) = 0 .$$

Hence, for all l and ε for which $2\varepsilon < \delta(z)$

$$\int \eta(x)\, e^{il\varepsilon(z,x)} \mu(dx) = 0 .$$

Choose l and ε so that $l\varepsilon = 1$. Then

$$\int \eta(x)\, e^{i(z,x)}\, \mu(dx) = 0$$

for all z. Let $H_1 = \{x : \eta(x) > 0\}$ and $H_2 = \{x : \eta(x) < 0\}$. Define the measures μ_1 and μ_2 by the equation

$$\mu_k(A) = (-1)^{k-1} \int_{H_k} \chi_A(x)\, \eta(x)\, \mu(dx) ,$$

where χ_A is the indicator of $A \in \mathfrak{B}$. Then (4) implies that

$$\int e^{i(z,x)}\, \mu_1(dx) = \int e^{i(z,x)}\, \mu_2(dx) .$$

Hence, because of the one-to-one relationship between characteristic functionals and measures, we will have $\mu_1(A) = \mu_2(A)$ for all $A \in \mathfrak{B}$. Since $\mu_k(A) = \mu_k(A \cap H_k)$ and $H_1 \cap H_2 = \phi$, we find that $\mu_1(A) = \mu_2(A) = 0$ for all A and $\eta(x) = 0 \pmod{\mu}$. □

§ 11. Orthogonal Systems of Polynomials

It makes sense to construct orthogonal systems of polynomials if an arbitrary measurable function can be decomposed in terms of them. Conditions under which this can be done are given in Theorem 1 of the previous section. Since the space of polynomials of degree $\leq n$ in a Hilbert space is infinite-dimensional, it is not convenient here to use the approach applicable in the case of finite-dimensional spaces for the construction of orthogonal polynomials. We will construct orthogonal spaces of polynomials (orthogonal in $L_2[\mu]$) in such a way that each space contains only polynomials of a given degree. Indeed, since $\tilde{\Pi}_n^2[\mu] \supset \tilde{\Pi}_{n-1}^2[\mu]$, we can define in $\tilde{\Pi}_n^2[\mu]$ the orthogonal complement to $\tilde{\Pi}_{n-1}^2[\mu]$ which we will denote by P_n. Then

$$\tilde{\Pi}_n^2[\mu] = P_n + \tilde{\Pi}_{n-1}^2[\mu] .$$

The intersection of P_n with $\Pi_{n-1}^2[\mu]$ contains only the point 0 so that all polynomials from P_n have degree n. P_0 are the polynomials of degree 0, i.e., the constants. Hence,

$$\tilde{\Pi}_n^2[\mu] = P_0 + P_1 + \cdots + P_n . \tag{1}$$

When $\bigcup \tilde{\Pi}_n^2[\mu]$ is dense in $L_2[\mu]$, for each function $\varphi \in L_2[\mu]$ one can find a sequence of polynomials $\pi_n \in P_n$ such that

$$\varphi(x) = \sum_{n=0}^{\infty} \pi_n(x) \,. \tag{2}$$

The system of subspaces P_n, $n = 0, 1, \ldots$ will be called an *orthogonal system of polynomials corresponding to the given measure* μ and decompositions of the form (2) are called *decompositions* w.r.t. this orthogonal system.

The construction of the subsets P_n is not entirely trivial if we do not start from already constructed subspaces $\tilde{\Pi}_n^2[\mu]$ (it is natural to assume that the subspaces P_n are simpler than $\tilde{\Pi}_n^2[\mu]$ and to first construct P_n and then $\tilde{\Pi}_n^2[\mu]$ with the help of (1)). One path which immediately suggests itself is to first construct the intersection of Π_0^n with P_n and then to close this intersection in $L_2[\mu]$, but this cannot be done if $\Pi_0^n \cap P_n$ is not dense in P_n.

We therefore choose another path: the inductive construction of the P_n, building at the same time a representation for functions from these spaces.

Let $\xi^{(n)} \in X^{0n}$ and $\eta(x) = \xi^{(n)}(x, \ldots, x)$. Then

$$\eta(x) = \pi_n(\xi^{(n)}, x) - \sum_{k=0}^{n-1} \varrho_k(\xi^{(n)}, x) \,, \tag{3}$$

where $\pi_n(\xi^{(n)}, x) \in P_n$ and $\varrho_k(\xi^{(n)}, x) \in P_k$. It is clear that the functions $\pi_n, \varrho_1, \ldots, \varrho_{n-1}$ appearing in (3) are uniquely determined by the form $\xi^{(n)}$ and depend linearly on it. Therefore, each function $\pi_n(x)$ is completely determined by the form of maximal degree (of degree n if such a form can be separated from $\pi_n(x)$).

Introduce in X^{0n} the scalar product

$$\langle \xi_1^{(n)}, \xi_2^{(n)} \rangle_n = \int \pi_n(\xi_1^{(n)}, x)\, \pi_n(\xi_2^{(n)}, x)\, \mu(dx) \,. \tag{4}$$

The completion of X^{0n} in this scalar product will be denoted by $\hat{X}^n$ and elements in $\hat{X}^n$ will be called generalized n-linear forms. They can be represented as sums

$$\sum \alpha_k \xi_k^{(n)} \,,$$

where $\xi_k^{(n)}$ is a complete orthonormalized (in the scalar product (4)) sequence of forms from X^{0k} and $\sum \alpha_k^2 < \infty$. We note that the correspondence

$$\xi^{(n)} \leftrightarrow \pi_n(\xi^{(n)}, x) \tag{5}$$

between X^{0n} and some subset of P_n is isometric due to (4) (the scalar product in P_n is generated by that in $L_2[\mu]$). Hence, this correspondence can be extended by continuity to all of $\hat{X}^n$. Since the polynomials $\pi_n(\xi^{(n)}, x)$ are dense in P_n (if this were not so then there would exist a function in P_n which would be orthogonal to all the polynomials in Π_0^n), Relation (5) can be extended to an isometric correspondence between $\hat{X}^n$ and P_n. We will denote forms in $\hat{X}^n$ by $\hat{\xi}^n$ and the polynomials from P_n corresponding to such forms by $\pi_n(\hat{\xi}^n, x)$.

It suffices to define $\pi_n(\hat{\xi}^n, x)$ for $\hat{\xi}^n \in X^{0n}$. Returning to (3), we see that $\varrho_k(\xi^{(n)}, x)$ can be represented as $\pi_k(\hat{\xi}^{(k)}, x)$, where $\hat{\xi}^{(k)} \in \hat{X}^k$. It is clear that $\hat{\xi}^{(k)}$ depends linearly on $\xi^{(n)}$. This means that there exists a linear operator $V_{n,k}$, mapping X^{0n} into $\hat{X}^k$, for which

$$\varrho_k(\xi^{(n)}, x) = \pi_k(V_{n,k}\,\xi^{(n)}, x)\,.$$

It follows from (3) that $\pi_n(\xi^{(n)}, x)$ can be represented as follows:

$$\pi_n(\xi^{(n)}, x) = \xi^{(n)}(x, \ldots, x) + \sum_{k=0}^{n-1} \pi_k(V_{n,k}\,\xi^{(n)}, x)\,. \tag{6}$$

Hence, if $\pi_k(\xi^{(k)}, x)$ is known for $k < n$, then to define $\pi_n(\xi^{(n)}, x)$ it is merely necessary to know the operators $V_{n,k}$. Then, in order to extend this definition to $\hat{X}^n$, it is still necessary to know the scalar product (4). For this scalar product we have the following recursive representation:

$$\begin{aligned}\langle\xi_1^{(n)}, \xi_2^{(n)}\rangle &= \int \xi_1^{(n)}(x, \ldots, x)\,\xi_2^{(n)}(x, \ldots, x)\,\mu(dx) + \\ &+ \int \xi_1^{(n)}(x, \ldots, x) \sum_{k=0}^{n-1} \pi_k(V_{n,k}\,\xi_2^{(n)}, x)\,\mu(dx) + \\ &+ \int \xi_2^{(n)}(x, \ldots, x) \sum_{k=0}^{n-1} \pi_k(V_{n,k}\,\xi_1^{(n)}, x)\,\mu(dx) + \sum_{k=0}^{n-1} \langle V_{n,k}\,\xi_1^{(n)}, V_{n,k}\,\xi_2^{(n)}\rangle_k\end{aligned}$$

and from (1) § 10

$$\langle\xi_1^{(n)}, \xi_2^{(n)}\rangle = \operatorname{tr} \xi_1^{(n)} \times \xi_2^{(n)} * \sigma_{2n} - \sum_{k=0}^{n-1} \langle V_{n,k}\,\xi_1^{(n)}, V_{n,k}\,\xi_2^{(n)}\rangle_k\,. \tag{7}$$

To determine $V_{n,k}$ we introduce a bilinear form $B_{n,k}$: for $\xi^{(n)} \in X^{0n}$ and $\xi^{(k)} \in X^{0k}$

$$B_{n,k}(\xi^{(n)}, \xi^{(k)}) = \int \xi^{(n)}(x, \ldots, x)\,\pi_k(\xi^{(k)}, x)\,\mu(dx)\,. \tag{8}$$

From (6) we obtain for $B_{n,k}$ the following recursion relation:

$$B_{n,k}(\xi^{(n)}, \xi^{(k)}) = \operatorname{tr} \xi^{(n)} \times \xi^{(k)} * \sigma_{n+k} + \sum_{j=0}^{k-1} B_{n,j}(\xi^{(n)}, V_{k,j}\,\xi^{(k)})\,. \tag{9}$$

Multiplying (6) by $\pi_k(\xi^{(k)}, x)$ and integrating w.r.t. μ we get

$$B_{n,k}(\xi^{(n)}, \xi^{(k)}) = -\langle V_{n,k}\,\xi^{(n)}, \xi^{(k)}\rangle_k\,. \tag{10}$$

From (10) $V_{n.k}$ is uniquely determined. Relations (7), (9) and (10) allow us to determine successively

$$\langle\cdot,\cdot\rangle_0,\ B_{10},\ V_{10},\ \langle\cdot,\cdot\rangle_1,\ B_{20},\ B_{21},\ V_{20},\ V_{21}\,,\ \text{and so forth.}$$

We construct as an example a system of orthogonal polynomials for the Gaussian measure μ with characteristic functional $\theta(z) = \exp\{-(A\,z, z)\}$, where A is some nuclear operator.

All of the formulas simplify considerably if we use the scalar product

$$(x, y)_+ = (A^{-1}\,x, y)$$

and calculate the traces in this product. In the scalar product $(\cdot\,,\cdot)_+$, the moment forms of μ are very simple: $\sigma_n^+ = 0$ for n odd and for n even

$$\sigma_n^+(z_1, \ldots, z_n) = \sum \prod_{k=1}^{n/2} (z_{i_k}, z_{j_k})_+\,,$$

where the sum is taken over all possible partitions of the numbers $1, 2, \ldots, n$ into $n/2$ pairs (i_k, j_k). The preceding formula follows from the equality

$$\sigma_n^+(z_1, \ldots, z_n) = i^{-n} \frac{\partial^n}{\partial\alpha_1 \cdots \partial\alpha_n} \int \exp\left\{i\left(x, \sum_1^n \alpha_k\, z_k\right)\right\} \mu(dx)\Bigg|_{\substack{\alpha_1=0\\ \vdots\\ \alpha_n=0}} =$$

$$= +\,i^{-n} \frac{\partial^n}{\partial\alpha_1 \cdots \partial\alpha_n} \exp\left\{-\frac{1}{2}\left(\sum_1^n \alpha_k\, z_k, \sum_1^n \alpha_k\, z_k\right)_+\right\}\Bigg|_{\substack{\alpha_1=0\\ \vdots\\ \alpha_n=0}}\,.$$

The traces in the scalar product $(\cdot\,,\cdot)_+$ will be denoted by tr_+.

We introduce a transformation of $\hat{X}^n$ into $\hat{X}^{n-2}$ defined by

$$\mathrm{tr}_+^2\, \xi^{(n)}(z_1, \ldots, z_{n-2}) = \sum_k \xi^{(n)}(e_k, e_k, z_1, \ldots, z_{n-2})\,,$$

where $\{e_k\}$ is some orthonormalized (in the scalar product $(\cdot\,,\cdot)_+$) basis, and seek the sequence of operators $V_{n.k}$ and scalar products $\langle\cdot\,,\cdot\rangle_n$. We evaluate

$$\mathrm{tr}_+\,(\xi^{(n)} \times \xi^{(k)} * \sigma_{n+k}^+)\,.$$

This expression is equal to zero when $n + k$ is odd. Assume $n + k = 2\,m$. We note that for the calculation of $\mathrm{tr}_+\, \xi^{(n+k)} * \sigma_{n+k}^+$, where $\xi^{(n+k)}$ is a linear form, it is necessary to break down the arguments of $\xi^{(n+k)}$ into all possible pairs, form the convolution of each pair (i.e., replace the pair of arguments by the same vectors from the orthonormalized basis and sum over the basis) and add the results. Assume the arguments of $\xi^{(n)} \times \xi^{(k)}$ have been broken down in such a way that there are s pairs which also contain the arguments of $\xi^{(n)}$ and $\xi^{(k)}$. Removing each form separately for the remaining pairs, we obtain

$$(\mathrm{tr}_+^2)^{\frac{n-s}{2}}\,\xi^{(n)} \qquad \text{and} \qquad (\mathrm{tr}_+^2)^{\frac{k-s}{2}}\,\xi^{(k)}\,.$$

There will be

$$C_n^s\, k\,(k-1)\cdots(k-s+1)\,(n-s+1)!!\,(k-s-1)!! = \\ = \frac{n!\,k!}{s!\,(n-s)!!\,(k-s)!!}$$ [1]

such decompositions.

Consequently, setting $\frac{k-s}{2} = j$ and $\frac{n-k}{2} = r$, we obtain

$$\mathrm{tr}_+\,(\xi^{(n)} \times \xi^{(k)} * \sigma_{n+k}^+) = \sum_{j \leq \frac{k}{2}} \frac{n!\,k!}{s!\,(n-s)!!\,(k-s)!!}\,\mathrm{tr}_+ \times \\ \times \{(\mathrm{tr}_+^2)^{j+r}\,\xi^{(n)} * (\mathrm{tr}_+^2)^j\,\xi^{(k)}\}\,.$$

Using this formula we can define

$$V_{2n,0}\,\xi^{(2n)} = -(2n-1)!!\,(\mathrm{tr}_+^2)^n\,\xi^{(2n)}$$

and

$$V_{2n+1,1}\,\xi^{(2n+1)} = -(2n+1)!!\,(\mathrm{tr}_+^2)\,\xi^{(2n+1)}\,.$$

Moreover,

$$B_{2n,2}(\xi^{(2n)}, \xi^{(2)}) = \frac{(2n)!}{(2n-2)!!}\,\mathrm{tr}_+\,((\mathrm{tr}_+^2)^{n-1}\,\xi^{(2n)} \times \xi^{(2)}) + \\ + (2n-1)!!\,(\mathrm{tr}_+^2)^n\,\xi^{(2n)}\,(\mathrm{tr}_+^2\,\xi^{(2)}) - \\ - B_{2n,0}(\xi^{(2n)}, V_{2,0}\,\xi^{(2)}) = \frac{(2n)!}{(2n-2)!!}\,\mathrm{tr}_+\,((\mathrm{tr}_+^2)\,\xi^{(2n)} \times \xi^{(2)})\,.$$

Consequently,

$$V_{2n,2}\,\xi^{(2n)} = -\frac{(2n)!}{(2n-2)!!\,2}\,(\mathrm{tr}_+^2)^{n-1}\,\xi^{(2n)}\,.$$

We can verify by induction that

$$B_{2n,2k}\,(\xi^{(2n)}, \xi^{(2k)}) = \frac{(2n)!}{(2n-2k)!!\,(2k)!}\,(\mathrm{tr}_+^2)^{n-k}\,\xi^{(2n)} \times \xi^{2(k)}$$

and

$$\langle \xi_1^{(2k)}, \xi_2^{(2k)} \rangle_{2k} = (2k)!\,\mathrm{tr}_+\,\xi_1^{(2k)} \times \xi_2^{(2k)}\,,$$

so that

$$V_{2n,2k}\,\xi^{(2n)} = -\frac{(2n)!}{(2n-2k)!!\,(2k)!}\,(\mathrm{tr}_+^2)^{n-k}\,\xi^{(2n)}\,.$$

---

[1] The symbol $n!!$ is used in Russian mathematical literature to denote the expression $1 \cdot 3 \cdots n$ or $2 \cdot 4 \cdots n$ depending on whether n is odd or even. — Transl.

In the same way, we get

$$V_{2n+1,2k+1}\,\xi^{(2n+1)} = -\frac{(2n+1)!}{(2n-2k)!!\,(2k+1)!}\,(\mathrm{tr}_+^2)^{n-k}\,\xi^{(2n+1)}$$

and finally

$$\langle \xi_1^{(n)}, \xi_2^{(n)} \rangle_n = n!\ \mathrm{tr}_+\, \xi_1^{(n)} \times \xi_2^{(n)}\,,$$

$$V_{n,k}\,\xi^{(n)} = \begin{cases} 0\,, & n+k \text{ odd;} \\ -\dfrac{n!}{(n-k)!!\,k!}\,(\mathrm{tr}_+^2)^{\frac{n-k}{2}}\,\xi^{(n)}\,, & n+k \text{ even.} \end{cases}$$

§ 12. Polynomials Orthogonal with Respect to a Weight Function

The constructions of the previous paragraph make sense only for measures from M_∞ and we can use the orthogonal polynomials we have constructed to decompose square-integrable functions for a still narrower class (see Theorem 1 § 10). In order to extend the class of measures for which it is possible to employ orthogonal polynomials, we can, as in finite-dimensional spaces, consider polynomials orthogonal w.r.t. a weight function. Let μ be a measure on $(X, \mathfrak{B})$ and $\varrho(x) \geq 0$ a $\mathfrak{B}$-measurable function such that for all $z \in X$ and $n \geq 0$

$$\int |(z, x)|^n\, \varrho(x)\, \mu(dx) < \infty\,. \tag{1}$$

We will consider polynomials $\pi_n(x)$ which are orthogonal w.r.t. the weight $\varrho(x)$. $\pi_n'(x)$ and $\pi_n''(x)$ are orthogonal if

$$\int \pi_n'(x)\, \pi_n''(x)\, \varrho(x)\, \mu(dx) = 0\,. \tag{2}$$

The construction of polynomials orthogonal w.r.t. ϱ is equivalent to the construction of a system of orthogonal functions of the form

$$\sqrt{\varrho(k)}\,\pi(x)\,, \tag{3}$$

where $\pi(x)$ is a polynomial. It is clear that such functions can be used to decompose measurable functions from $L_2[\mu]$. By suitably choosing $\varrho(x)$, it is possible to provide these functions with sufficiently simple analytic properties. To construct a system of orthogonal functions of the form (3) we can use the results of the previous section.

Introduce a new measure ν defined by

$$\nu(A) = \int_A \varrho(x)\, \mu(dx)\,. \tag{4}$$

Then (1) and (2) are rewritten as

$$\int |(z, x)|^n\, \nu(dx) < \infty, \quad z \in X, \quad n > 0; \quad \int \pi_n'(x)\, \pi_n''(x)\, \nu(dx) = 0\,.$$

Thus, the measure ν will belong to M_∞ and the system of polynomials will be orthogonal w.r.t. ν.

As shown in § 11, it is sufficient to construct the system of orthogonal subspaces $P_n[\nu]$ of ν-measurable polynomials of degree n. As functions of the form (3) we can take elements from $P_n[\nu]$ multiplied by $\sqrt{\varrho(x)}$. To obtain this sequence of orthogonal functions it is still necessary to construct bases in each subspace $P_n[\nu]$ (these bases are constructed by orthogonalizing an arbitrary sequence whose linear closure is dense in $P_n[\nu]$).

The main difficulty arising in connection with the construction of orthogonal polynomials w.r.t. the weight ϱ is the calculation of the moment functions of the measure ν according to the characteristics of the measure μ in terms of which it is given. The moment functions of ν have the form

$$\sigma_n^{(\nu)}(z_1, \ldots, z_n) = \int (z_1, x) \cdots (z_n, x)\, \varrho(x)\, \mu(dx)\,. \tag{5}$$

We remark that the task of finding suitable procedures for the calculation of integrals w.r.t. μ for various classes of functions is nontrivial so that it is desirable to simplify it as much as possible. It is quite convenient for the calculation of the moment functions (5) to find the characteristic functional of the measure ν from which these functions can then be found by means of differentiation.

We now consider a concrete case of a function $\varrho(x)$ for which the characteristic functional of the measure μ can be conveniently represented. Let

$$\varrho(x) = \exp\{-(x, x)\} \tag{6}$$

(this weight is analogous to that for the Čebyšev-Hermite polynomials in the one-dimensional case). Note that the function

$$|(z, x)|^n\, e^{-(x,x)} \leq |z|^n\, |x|^n\, e^{-|x|^2}$$

is bounded for arbitrary $z \in X$ and $n > 0$. Hence, the integral (1) exists. Since for all $z \in X$

$$\int e^{t(z,x)}\, e^{-(x,x)}\, \mu(dx)$$

is an entire analytic function of t due to the convergence of the integral

$$\int e^{|t|\cdot|z|\cdot|x| - |x|^2}\, \mu(dx)$$

(the function $|t| \cdot |z| \cdot |x| - |x|^2$ is bounded above by the number $\frac{1}{2}\, |t|^2\, |z|^2$), the conditions of Theorem 1 § 10 hold for ν for any measure μ.

To calculate the characteristic function $\theta_\nu(z)$ of the measure ν given in (4) with $\varrho(x)$ as in (6), we note that

$$e^{-(P_L x, P_L x)} = \int e^{i(x,u)}\, \lambda^L(du)\,,$$

where λ^L is the Gaussian measure with mean zero and correlation operator P_L (here L is a finite-dimensional subspace of X, and P_L is projection onto this subspace). Hence,

$$\begin{aligned}\theta_\nu(z) &= \lim_{L\uparrow X} \int e^{i(z,x)} \int e^{i(x,u)} \lambda^L(du)\, \mu(dx) = \\ &= \lim_{L\uparrow X} \int \left[\int e^{i(z+u,x)} \mu(dx)\right] \lambda^L(du) \\ &= \lim_{L\uparrow X} \int \theta_\mu(z+u)\, \lambda^L(du)\,,\end{aligned}$$

where $\lim_{L\uparrow X}$ denotes the limit w.r.t. an increasing sequence of finite-dimensional subspaces L_n for which $\bigcup_n L_n$ is dense in X, and θ_μ is the characteristic functional of the measure μ. Let λ_L be the projection of λ^L onto L. Then

$$\lambda_L(A) = (4\pi)^{-m/2} \int e^{-(x,x)/4}\, m_L(dx)\,.$$

The $\{\lambda_L(A)\}$ form a compatible family of finite-dimensional distributions (see the example in § 2), to which there corresponds some weak distribution. Since θ_μ is the characteristic functional of a measure, on the basis of the result in § 4 there exists a nuclear operator C such that $\theta(z)$ is continuous in the scalar product $(Cz, z) = (z, z)_-$. Hence, $\theta_\mu(z+u)$, as a function of u, can be extended by continuity to X_-, the completion of X in the scalar product $(.,.)_-$. But on the space X_- there exists a measure $\bar{\lambda}$ with finite-dimensional distributions $\{\lambda_L\}$ and this measure is Gaussian. Clearly,

$$\lim_{L\uparrow X} \int \theta_\mu(z+u)\, \lambda^L(du) = \int \theta_\mu(z+u)\, \bar{\lambda}(du)$$

(we denote the extension of $\theta_\mu(z+u)$ to X_- by the same symbol). In order to avoid going into the extension of X we can write

$$\theta_\nu(z) = \int \theta_\mu(z+u)\, \lambda_*(du)\,. \tag{7}$$

This formula is convenient because of the fact that to determine $\theta_\nu(z)$ we need only integrate $\theta_\mu(z)$ w.r.t. a single measure.

A system of orthogonal functions in $L_2[\mu]$ can also be constructed by the usual method of taking an arbitrary sequence of functions from $L_2[\mu]$ and orthogonalizing them by means of the Gram-Schmidt process. To employ such a method one must be able to compute the integral

$$\int \varphi_n(x)\, \varphi_m(x)\, \mu(dx)\,, \tag{8}$$

where $\{\varphi_n\}$ is the sequence under consideration. As already mentioned, the evaluation of these integrals can, generally speaking, cause some difficulty. However, one can define a class of functions $\varphi_n(x)$ for which integrals of the form (8) can be quite easily calculated provided the characteristic functional of the measure μ is given. These are the *trigonometric functions.*

Here it is convenient to consider complex-valued $\mathfrak{B}$-measurable functions $\varphi(x)$ for which

$$\int |\varphi(x)|^2 \mu(dx) < \infty .$$

The space of such functions will be denoted by $\bar{L}_2[\mu]$. The scalar product in $\bar{L}_2[\mu]$ is naturally defined as

$$(\varphi_1, \varphi_2) = \int \varphi_1(x) \bar{\varphi}_2(x) \mu(dx) . \tag{9}$$

Trigonometric functions will be understood as functions of the form $\exp\{i(z, x)\}$ for all possible $z \in X$ (the argument of the function is taken as x). Linear combinations of trigonometric functions are called *trigonometric polynomials.* It is clear that such polynomials are dense in the sense of convergence in $\bar{L}_2[\mu]$ in the set of bounded cylinder functions which in turn is dense in the set of bounded continuous functions (for which functions $\varphi(x) = \lim_{L \uparrow X} \varphi(P_L x)$ holds); finally, the last set is everywhere dense in $\bar{L}_2[\mu]$. Hence, using the set of functions $\exp\{i(z, x)\}$, where z runs through some denumerable set dense in X, we can construct a complete system of trigonometric polynomials. The values of the integrals

$$\int \exp\{i(z_1, x)\} \overline{\exp\{i(z_2, x)\}} \mu(dx) =$$
$$= \int \exp\{i(z_1 - z_2, x)\} \mu(dx) = \theta_\mu(z_1 - z_2)$$

are known if the characteristic functional of the measure μ is known. When μ has a representation of the form

$$\mu = \prod_{k=1}^{\infty} \mu_{L(k)} ,$$

where the $\mu_{L(k)}$ are measures on one-dimensional subspaces generated by vectors e_k forming an orthonormalized basis in X, then the construction of an orthogonal system of trigonometric polynomials is much easier. Let $\psi_{k,n}((x, e_k))$ be a complete system of trigonometric polynomials orthogonal w.r.t. $\mu_{L(k)}$. Then the trigonometric polynomials of the form

$$\prod_{k=1}^{m} \psi_{k, n_k}((x, e_k))$$

generate for all possible m and $n_1, \ldots, n_m$ a complete orthogonal system of trigonometric polynomials w.r.t. μ in such a way that for such measures the problem is reduced to the construction of an orthogonal system of trigonometric polynomials for measures in one-dimensional spaces.

Chapter 3. Absolute Continuity of Measures

§ 13. The Radon-Nikodym Theorem. Conditional Measures

Absolute continuity and singularity play a very important role in the study of measures in infinite-dimensional spaces, for example, Hilbert space. Although there can be no theory treating such questions for finite-dimensional spaces which of great interest, such a theory for infinite-dimensional spaces is possible. It contains such topics as the investigation of the absolute continuity and singularity of various concrete classes of measures, the finding of general conditions for absolute continuity or singularity in terms of finite-dimensional distributions, and other characteristics defining the measures. An important problem is the calculation of the density of a measure w.r.t. another when the measures are absolutely continuous and the determination of the non-overlapping sets on which singular measures are concentrated.

We recall some basic definitions. Let two measures μ and ν be given on a measurable space $(X, \mathfrak{B})$. We say that ν is *absolutely continuous w.r.t.* μ if $\nu(A) = 0$ whenever $\mu(A) = 0$, $A \in \mathfrak{B}$. The absolute continuity of ν w.r.t. μ is denoted by the symbolism $\nu \ll \mu$. If $\nu \ll \mu$ and $\mu \ll \nu$ then we write $\nu \sim \mu$ and say that ν and μ are *equivalent.* We call μ and ν *singular* if there is a set $F \subset X$ such that $\mu(X - F) = 0$ and $\nu(F) = 0$. Singularity of μ and ν is denoted as: $\mu \perp \nu$ (we also say that μ is *orthogonal* to ν).

For the absolute continuity of a measure ν w.r.t. μ it is necessary and sufficient that for any $\varepsilon > 0$ there exist a $\delta > 0$ such that $\nu(A) < \delta$ when $\mu(A) < \varepsilon$. It is even sufficient that this condition hold for A from some algebra $\mathfrak{A}_0$ whose monotone closure coincides with $\mathfrak{B}$.

A fundamental feature of the absolute continuity of measures is contained in the

Radon-Nikodym Theorem. *In order for a measure ν to be absolutely continuous w.r.t. a measure μ, it is necessary and sufficient that there exist a $\mathfrak{B}$-measurable function $\varrho(x)$, integrable w.r.t. μ, such that for* *all* $A \in \mathfrak{B}$

$$\nu(A) = \int_A \varrho(x)\, \mu(dx)\,. \tag{1}$$

The function $\varrho(x)$ is called the *density* (or *derivative*) of the measure ν w.r.t. the measure μ and is written as

$$\varrho(x) = \frac{d\nu}{d\mu}(x)\,.$$

If ν and μ are arbitrary measures, then ν can be represented in the form $\nu_1 + \nu_2$,where $\nu_1 \ll \mu$ and $\nu_2 \perp \mu$. ν_1 is called the *absolutely continuous component* of μ and ν_2 the *singular component*. The density $\varrho_1(x)$ of the absolutely continuous component w.r.t. the measure μ is also called the density of ν w.r.t. μ and is written

$$\varrho_1(x) = \frac{d\nu}{d\mu}(x)\,.$$

Hence, the condition for singularity of ν w.r.t. μ can be written as:

$$\frac{d\nu}{d\mu} = 0 \qquad (\text{mod}\ \mu)\,.$$

Let $\nu \ll \mu$ and $\frac{d\nu}{d\mu}(x) > 0$ (mod μ). Then $\mu \ll \nu$ and

$$\frac{d\mu}{d\nu}(x) = \left[\frac{d\nu}{d\mu}(x)\right]^{-1}.$$

Indeed, for a $\mathfrak{B}$-measurable nonnegative function f

$$\int f(x)\, \mu(dx) = \int f(x) \left[\frac{d\nu}{d\mu}(x)\right]^{-1} \frac{d\nu}{d\mu}(x)\, \mu(dx) = \int f(x) \left[\frac{d\nu}{d\mu}(x)\right]^{-1} \nu(dx)\,,$$

provided the integral on the left exists.

It is also easy to verify the following useful relation: for $\mu \ll \nu \ll \pi$

$$\frac{d\mu}{d\pi}(x) = \frac{d\mu}{d\nu}(x)\,\frac{d\nu}{d\pi}(x)\,.$$

We will often use the following elementary fact. Let f be a map of the measurable space $(X, \mathfrak{B})$ into $(Y, \mathfrak{C})$ Assume the measure $\tilde{\mu}$ and $\tilde{\nu}$ on $(Y, \mathfrak{C})$ are the images of the measures μ and ν under the map f: for $C \in \mathfrak{C}$

$$\tilde{\mu}(C) = \mu(f^{-1}(C))\,, \qquad \tilde{\nu}(C) = \nu(f^{-1}(C))\,.$$

Then from $\nu \ll \mu$ $(\nu \perp \mu)$ follows $\tilde{\nu} \ll \tilde{\mu}$ $(\tilde{\nu} \perp \tilde{\mu})$.

The Radon-Nikodym theorem finds application in all branches of analysis. In measure theory itself, it can be used to construct conditional measures. This notion will be employed in the sequel and hence we will dwell on it in more detail.

Let $(X, \mathfrak{B})$ be a measurable Hilbert space. Take some σ-algebra $\mathfrak{B}_0$ of subsets of the algebra $\mathfrak{B}$. Let μ be a measure on $(X, \mathfrak{B})$. Its restriction to $\mathfrak{B}_0$ will also be denoted by μ. For any $A \in \mathfrak{B}$, the measure on $\mathfrak{B}_0$ given by the equation $\mu(A \cap A_0) = \mu_A(A_0)$ is absolutely continuous w.r.t. the measure μ on $\mathfrak{B}_0$ since

$$\mu_A(A_0) \leq \mu(A_0), \quad A_0 \in \mathfrak{B}_0.$$

By the Radon-Nikodym theorem, the density of the measure $\mu_A(A_0)$ w.r.t. $\mu(A_0)$ exists. We denote it by $\mu(A/x)$:

$$\mu(A \cap A_0) = \mu_A(A_0) = \int_{A_0} \mu(A/x)\, \mu(dx), \quad A_0 \in \mathfrak{B}_0. \tag{2}$$

The function $\mu(A/x)$ is $\mathfrak{B}_0$-measurable and nonnegative. Since

$$\mu(A \cap \{x: \mu(A/x) > 1\}) = \int_{\{x:\mu(A/x)>1\}} \mu(A/x)\, \mu(dx),$$

$\mu(\{x: \mu(A/x) > 1\}) = 0$ so that $\mu(A/x) \leq 1 \pmod{\mu}$. Note that $\mu(A/x)$ is uniquely $(\bmod\, \mu)$ defined as a $\mathfrak{B}_0$-measurable function, satisfying (2) for all $A_0 \in \mathfrak{B}_0$. Hence, for any sequence of non-overlapping sets $A_k \in \mathfrak{B}$, it follows from the equality

$$\int_{A_0} \mu\Big(\bigcup_k A_k/x\Big)\mu(dx) = \mu\Big(\bigcup_k A_k \cap A_0\Big) = \sum_k \int_{A_0} \mu(A_k/x)\, \mu(dx) =$$
$$= \int_{A_0} \sum_k \mu(A_k/x)\, \mu(dx)$$

that

$$\mu\Big(\bigcup_k A_k/x\Big) = \sum_k \mu(A_k/x) \qquad (\bmod\, \mu).$$

Hence, $\mu(A/x)$ possesses the properties of a measure as a function of x for the measure μ. It does not follow from this that $\mu(A/x)$ will be a measure on A for fixed x. We will prove that one can choose a function $\mu(A/x)$ (for each A its values can be altered on a set of measure 0) such that $\mu(A/x)$ will be a measure on A for almost all x (w.r.t. μ). Such a function $\mu(A/x)$, i.e., a function satisfying the following conditions: a) $\mu(A/x)$ is $\mathfrak{B}_0$-measurable in x, b) for all $A_0 \in \mathfrak{B}_0$ it satisfies (2) and c) $\mu(A/x)$ is a measure on A for almost all x (w.r.t. μ), will be called a *conditional measure* for μ w.r.t. the σ-algebra $\mathfrak{B}_0$.

To prove the existence of a conditional measure, we use the characteristic functional. Define a $\mathfrak{B}_0$-measurable function $\theta(z/x)$ by means of

$$\int_{A_0} e^{i(z,x)}\, \mu(dx) = \int_{A_0} \theta(z/x)\, \mu(dx) \tag{3}$$

which is to hold for all $A_0 \in \mathfrak{B}_0$. $\theta(z/x)$ can be represented as $\theta_1(z/x) + i\,\theta_2(z/x)$, where

$$\int_{A_0} (1 - \cos(z, x))\, \mu(dx) = \int_{A_0} (1 - \theta_1(z/x))\, \mu(dx)\,,$$

$$\int_{A_0} (1 - \sin(z, x))\, \mu(dx) = \int_{A_0} (1 - \theta_2(z/x))\, \mu(dx)\,,$$

and $\theta_1(z/x)$ and $\theta_2(z/x)$ are measurable functions whose existence is guaranteed by the Radon-Nikodym theorem.

Our goal is to prove that $\theta(z/x)$ can be so defined that for almost all x it will be a characteristic functional.

Let $\varphi(x)$ be a continuous bounded function. We define the integral

$$\int \varphi(y)\, \mu(dy/x) \tag{4}$$

as a $\mathfrak{B}_0$-measurable function such that for all $A_0 \in \mathfrak{B}_0$

$$\int_{A_0} \int \varphi(y)\, \mu(dy/x)\, \mu(dx) = \int_{A_0} \varphi(y)\, \mu(dy)\,. \tag{5}$$

We remark that this integral is the limit w.r.t. μ of the integral sums

$$\sum_k \varphi(y_k)\, \pi(A_k/x)\,,$$

for $\lambda \to 0$, where the A_k are pairwise disjoint, $\bigcup_k A_k = X$, $y_k \in A_k$, and $\lambda = \sup_k \left[\sup_{y \in A_k} \varphi(y) - \inf_{y \in A_k} \varphi(y)\right]$.

We choose a denumerable algebra of sets $\mathfrak{A}_0$ such that for any $\varepsilon > 0$ a sequence of sets $A_k \in \mathfrak{A}_0$ can be found for which $\bigcup_k A_k = X$ and the diameters of the A_k do not exceed ε.

Select in $\mathfrak{A}_0$ a system of sets $A_{n,k}$ such that $A_{n,k}$ and $A_{n,j}$ do not overlap for $k \neq j$,

$$\bigcup_{k=1}^{\infty} A_{n,k} = X\,,$$

the diameter of $A_{n,k}$ does not exceed $1/n$, and such that for each k there exists a j with $A_{n+1,k} \subset A_{n,j}$.

Denote by Y the set of all x for which $\mu(A/x)$ is finitely additive on $\mathfrak{A}_0$ and for all n and k

$$\pi(A_{n,k}/x) = \sum_j \mu\,(A_{n+1,j} \cap A_{n,k}/x)\,.$$

Since there are only denumerably many additivity conditions and each such condition is satisfied almost everywhere, $\mu(Y) = 1$. If $\varphi(x)$ is an arbitrary uniformly continuous bounded function on X, then

$$\int \varphi(y)\, \mu(dy/x)$$

can be defined as the limit

$$\lim_{n\to\infty} \sum_{k=1}^{\infty} \mu(A_{n,k}/x)\, \varphi(y_{n,k}), \tag{6}$$

where $y_{n,k} \in A_{n,k}$. This limit exists for all $x \in Y$. It is easy to see that

$$\theta(z/x) = \int e^{i(z,x)}\, \mu(dy/x) \tag{7}$$

and since $e^{i(z,x)}$ is a uniformly continuous function, $\theta(z/x)$ can be calculated with the aid of (6). Since

$$\lim_{n\to\infty} \int \left(1 - e^{-\frac{1}{n}(x,x)}\right) \mu(dx) = 0,$$

if we set

$$B_n = \left\{x: \int \left(1 - e^{-\frac{1}{n}(y,y)}\right) \mu(dy/x) > \left[\int \left(1 - e^{-\frac{1}{n}(x,x)}\right) \mu(dx)\right]^{1/2}\right\},$$

then we get

$$\int \left(1 - e^{-\frac{1}{n}(x,x)}\right) \mu(dx) \geq \int_{B_n}\int \left(1 - e^{-\frac{1}{n}(y,y)}\right) \mu(dy/x)\, \mu(dx) \geq$$

$$\geq \left[\int \left(1 - e^{-\frac{1}{n}(x,x)}\right) \mu(dx)\right]^{1/2} \mu(B_n),$$

so that $\mu(B_n) \to 0$ for $n \to \infty$. If $Y_1 = X - \bigcup_{n=1}^{\infty} \bigcup_{l=n}^{\infty} B_l$, then $\mu(Y_1) = 1$ and for all $x \in Y \cap Y_1$

$$\lim_{n\to\infty} \int \left(1 - e^{-\frac{1}{n}(y,y)}\right) \mu(dy/x) = 0,$$

since on Y the integral following the limit sign depends monotonically on n and for $x \in Y_1$ one can choose n large enough so that $x \notin B_n$, which means that

$$\int \left(1 - e^{-\frac{1}{n}(y,y)}\right) \mu(dy/x) \leq \left[\int \left(1 - e^{-\frac{1}{n}(x,x)}\right) \mu(dx)\right]^{1/n}.$$

For all $x \in Y \cap Y_1$, $\theta(z/x)$ — defined as a limit of the form (6) — is a function depending continuously on z. In fact,

$$\left| \int e^{i(z_1,y)}\, \mu(dy/x) - \int e^{i(z_2,y)}\, \mu(dy/x) \right| \leq$$

$$\leq \int |1 - e^{i(z_1 - z_2, y)}|\, \mu(dy/x) \leq$$

$$\leq \int |1 - e^{i(z_1-z_2,y)}|\, e^{-\frac{1}{n}(y,y)}\, \mu(dy/x) + 2\int \left(1 - e^{-\frac{1}{n}(y,y)}\right) \mu(dy/x) \leq$$

$$\leq |z_1 - z_2| \int |y|\, e^{-\frac{1}{n}(y,y)}\, \mu(dy/x) + 2\int \left(1 - e^{-\frac{1}{n}(y,y)}\right) \mu(dy/x).$$

The second integral can be made arbitrarily small by suitable choice of n and the first by suitable choice of $|z_1 - z_2|$. The continuity of $\theta(z/x)$ is proved. Since

$$\sum_{k,j=1}^{n} \theta(z_k - z_j/x)\, \alpha_k \bar{\alpha}_j = \int |\sum e^{i(z_k, y)} \alpha_k|^2 \mu(dy/x),$$

$\theta(z/x)$ for $x \in Y_1 \cap Y$ is positive-definite in z. Finally, for large enough n and $x \in Y_1 \cap Y$

$$\mathrm{Re}\,(1 - \theta(z/x)) = \int (1 - \cos(z, y))\, \mu(dy/x) \leq$$

$$\leq \frac{1}{2} \int e^{-\frac{1}{n}|y|^2} (1 - \cos(z, y))\, \mu(dy/x) + 2 \int \left(1 - e^{-\frac{1}{n}(y,y)}\right) \mu(dy/x) \leq$$

$$\leq \frac{\varepsilon}{2} + \frac{1}{2} \int (z, y)^2 e^{-\frac{1}{n}|y|^2} \mu(dy/x).$$

It's easy to see that the expression

$$C(z_1, z_2) = \frac{1}{2} \int (z_1, y)\,(z_2, y)\, e^{-\frac{1}{n}|y|^2} \mu(dy/x), \quad x \in Y \cap Y_1,$$

is a positive-definite, symmetric bilinear form and for an arbitrary orthonormalized basis $\{e_k\}$

$$\sum_{1}^{l} C(e_k, e_k) \leq \frac{1}{2} \int |y|^2 e^{-\frac{1}{n}|y|^2} \mu(dy/x).$$

Hence, there exists a linear operator D such that $C(z_1, z_2) = (D z_1, z_2)$. Thus, for all $x \in Y_1 \cap Y$, $\theta(z/x)$ satisfies the hypotheses of the Minlos-Sazonov theorem so that there exists a family of measures $\bar{\mu}(dy/x)$ for which

$$\theta(z/x) = \int e^{i(z,y)} \bar{\mu}(dy/x).$$

Using the $\mathfrak{B}_0$-measurability of $\theta(z/x)$, it is easy to convince oneself that

$$\int \varphi(y)\, \bar{\mu}(dy/x) \tag{8}$$

is $\mathfrak{B}_0$-measurable for all continuous bounded functions φ (it is necessary to approximate $\varphi(y)$ by cylinder functions $\varphi(P_L y)$, and then approximate the cylinder functions by trigonometric polynomials). Since with the help of a limit passage one can obtain all bounded $\mathfrak{B}$-measurable functions from bounded continuous functions, (8) will be a $\mathfrak{B}_0$-measurable function of x for all bounded $\mathfrak{B}$-measurable functions. Hence, $\bar{\mu}(B/x)$ is $\mathfrak{B}$-measurable for all $B \in \mathfrak{B}$. Starting from the equality

$$\int_{A_0} \left[\int e^{i(z,y)} \bar{\mu}(dy/x)\right] \mu(dy) = \int_{A_0} e^{i(z,y)} \bar{\mu}_{A_0}(dy),$$

where $\bar{\mu}_{A_0}(B) = \mu\,(A_0 \cap B)$ is some measure on $\mathfrak{B}$, we can show that

$$\int e^{i(z,y)}\,\bar{\mu}_{A_0}(dy) = \int_{A_0} \theta(z/x)\,\mu(dx) = \int_{A_0} e^{i(z,y)}\,\mu(dy)\,,$$

so that $\bar{\mu}_{A_0}(B) = \mu\,(A_0 \cap B)$. In other words, $\bar{\mu}(B/x)$ also satisfies

$$\mu\,(A_0 \cap B) = \int_{A_0} \bar{\mu}(B/x)\,\mu(dx) \tag{9}$$

as does $\mu(B/x)$, so that $\bar{\mu}(B/x)$ is a conditional measure for μ w.r.t. the σ-algebra $\mathfrak{B}_0$. This proves the existence of a conditional measure.

Quite important is the case in which the σ-algebra $\mathfrak{B}_0$ is generated by some finite or infinite set of functions $\{\varphi_\alpha\}$, i.e., when it is the minimal σ-algebra w.r.t. which all these functions are measurable. We remark finally that (9) is equivalent to the following assertion: for any bounded $\mathfrak{B}_0$-measurable function $\psi(x)$ and $\mathfrak{B}$-measurable function $\varphi(x)$ for which $\int |\varphi(x)|\,\mu(dx) < \infty$

$$\int \psi(x)\,\varphi(x)\,\mu(dx) = \int \psi(x) \int \varphi(y)\,\bar{\mu}(dy/x)\,\mu(dx)\,. \tag{10}$$

Starting from (10) one can establish the fact that for any bounded $\mathfrak{B}_0$-measurable function $\xi(x)$ and a function φ for which $\int |\varphi(x)| \times \mu(dx) < \infty$, one has

$$\int \xi(y)\,\varphi(y)\,\bar{\mu}(dy/x) = \xi(x) \int \varphi(y)\,\bar{\mu}(dy/x) \qquad (\mathrm{mod}\,\mu)\,. \tag{11}$$

§ 14. Martingales and Semi-Martingales

For the sake of brevity we will denote the measurable space $(X, \mathfrak{B})$ on which a measure μ is defined, by $(X, \mathfrak{B}, \mu)$. The triplet $(X, \mathfrak{B}, \mu)$ will be called *a space with measure*. A sequence of measurable functions $\{\varphi_n(x),\ n = 1, 2, \ldots\}$ on $(X, \mathfrak{B}, \mu)$ is called a *martingale* if for each n

$$\int |\varphi_n(x)|\,\mu(dx) < \infty$$

and for each $\mathfrak{A}_n$-measurable bounded function $\psi(x) \geq 0$, where $\mathfrak{A}_n$ is the minimal σ-algebra w.r.t. which $\varphi_1, \ldots, \varphi_n$ are measurable, we have

$$\int \varphi_{n+1}(x)\,\psi(x)\,\mu(dx) = \int \varphi_n(x)\,\psi(x)\,\mu(dx)\,. \tag{1}$$

If $\geq$ holds in place of equality in (1), then the sequence is called a *semi-martingale*[1]; if $\leq$ holds, then we will call it a *sub-martingale*. It is easy to show that (1) holds for all nonnegative $\mathfrak{A}_n$-measurable functions $\psi(x)$ for which

$$\int (|\varphi_n(x)| + |\varphi_{n+1}(x)|)\,\psi(x)\,\mu(dx)$$

[1] Some authors prefer the term "super-martingale". — Transl.

exists. Functions meeting this condition satisfy (1) for semi-martingales when the sign there is $\geq$ and for sub-martingales when it is $\leq$.

All of these sequences are distinguished by the fact that under very general restrictions,

$$\lim_{n\to\infty} \varphi_n(x) \tag{2}$$

exists almost everywhere (mod μ). Such a restriction on φ_n is simple to give: it is necessary that

$$\sup_n \int |\varphi_n(x)| \, \mu(dx) < \infty \, . \tag{3}$$

The main purpose of this section will be to prove the existence of the limit (2) under the restriction (3) for martingales, semi-martingales and sub-martingales.

It turns out that two other sequences can be rather easily expressed by means of martingales.

Lemma 1. *For any semi-martingale (sub-martingale) φ_n there exists a sequence of functions $\zeta_n(x)$ such that* a) *$\zeta_n(x)$ is $\mathfrak{A}_n$-measurable,* b) *$\varphi_n - \zeta_n$ is a martingale and* c) *$\zeta_n(x)$ increases with n (decreases with n) almost everywhere* (w.r.t. μ).

Proof. Set

$$\zeta_{n+1}(x) - \zeta_n(x) = \int \varphi_{n+1}(y) \, \mu(dy, \mathfrak{A}_n/x) - \varphi_n(x) \, , \tag{4}$$

where $\mu(B, \mathfrak{A}_n/x)$ is a conditional measure for μ w.r.t. the σ-algebra $\mathfrak{A}_n$. The existence of such a conditional measure was proved in § 13. Clearly, $\zeta_{n+1}(x) - \zeta_n(x)$ is an $\mathfrak{A}_n$-measurable function. Multiplying this function by an arbitrary bounded nonnegative function $\psi(x)$ measurable w.r.t. $\mathfrak{A}_n$, we find on the basis of Eq. (10) § 13 that

$$\begin{aligned} &\int [\zeta_{n+1}(x) - \zeta_n(x)] \, \psi(x) \, \mu(dx) = \\ &= \int \int \varphi_{n+1}(y) \, \mu(dy, \mathfrak{A}_n/x) \, \psi(x) \, \mu(dx) - \int \varphi_n(x) \, \psi(x) \, \mu(dx) = \\ &= \int \varphi_{n+1}(x) \, \psi(x) \, \mu(dx) - \int \varphi_n(x) \, \psi(x) \, \mu(dx) \, . \end{aligned} \tag{5}$$

This implies, if we set

$$\zeta_{n+1}(x) = \sum_{k=1}^{n} [\xi_{k+1}(x) - \xi_k(x)] \, , \qquad \xi_1(x) = 0 \, ,$$

that

$\int [\zeta_{n+1}(x) + \varphi_{n+1}(x)] \, \psi(x) \, \mu(dx) = \int [\zeta_n(x) + \varphi_n(x)] \, \psi(x) \, \mu(dx)$. Hence, for this $\zeta_n(x)$, a) and b) of the lemma are correct. Set $\psi(x) = 1$ in (5) if $\zeta_{n+1}(x) - \zeta_n(x) < 0$ and $\psi(x) = 0$ otherwise. Then, using the definition of a semi-martingale, we get

$$-\int \psi(x) \, \mu(dx) = \int [\zeta_{n+1}(x) - \zeta_n(x)] \, \psi(x) \, \mu(dx) \geq 0 \, .$$

This means that $\zeta_{n+1}(x) \geq \zeta_n(x)$ a.e. (w.r.t. μ). The reversed inequality for sub-martingales is established analogously.

Remark. In the course of proving the lemma we showed that for a semi-martingale $\{\varphi_n(x)\}$, the inequality

$$\int \varphi_{n+1}(y)\, \mu(dy, \mathfrak{A}_n/x) \geq \varphi_n(x) .$$

holds for all n. For martingales, equality holds here. It's easy to show that when this condition is satisfied, $\{\varphi_n(x)\}$ will be a semi-martingale if, for all n,

$$\int |\varphi_n(x)|\, \mu(dx) < \infty .$$

Let $\zeta_n(x)$ be constructed as in Lemma 1. Then for a semi-martingale

$$\int \zeta_n(x)\, \mu(dx) = \int [\zeta_n(x) - \varphi_n(x)]\, \mu(dx) + \int \varphi_n(x)\, \mu(dx) .$$

Taking $\psi(x) = 1$ in (1), we find that for a martingale $\{\varphi_n(x)\}$

$$\int \varphi_n(x)\, \mu(dx) = \int \varphi_1(x)\, \mu(dx) .$$

Hence,

$$\int \zeta_n(x)\, \mu(dx) = \int [\zeta_1(x) - \varphi_1(x)]\, \mu(dx) + \int \varphi_n(x)\, \mu(dx) = \\ = \int \varphi_n(x)\, \mu(dx) - \int \varphi_1(x)\, \mu(dx) .$$

If (3) holds for a semi-martingale $\varphi_n(x)$, then

$$\sup_n \int \zeta_n(x)\, \mu(dx) < \infty$$

so that the nondecreasing sequence $\zeta_n(x)$ has a limit a.e. and this limit is also integrable (w.r.t. μ). In the case of a sub-martingale $\varphi_n(x)$, $-\varphi_n(x)$ will be a semi-martingale and then the sequence $-\zeta_n(x)$ will have an integrable limit.

Thus, to prove the existence of the limit (2) for each of the three sequences defined in this section it is sufficient to merely consider martingales.

Lemma 2. *If $\{\varphi_n(x)\}$ is a martingale, then for all n and $a > 0$*

$$\mu\left(\left\{x\colon \sup_{k \leq n} \varphi_k(x) \geq a\right\}\right) \leq \frac{1}{a} \sup_n \int \varphi_n^+(x)\, \mu(dx) , \tag{6}$$

where $\varphi_n^+(x) = \varphi_n(x)$ for $\varphi_n(x) \geq 0$ and $\varphi_n^+(x) = 0$ for $\varphi_n(x) < 0$. If $\sup_n \int \varphi_n^+(x)\, \mu(dx) < \infty$, then

$$\mu\left(\left\{x\colon \sup_{k} \varphi_k(x) \geq a\right\}\right) \leq \frac{1}{a} \sup_n \int \varphi_n^+(x)\, \mu(dx) . \tag{7}$$

Proof. Let $\chi_k(x) = 1$ if $\varphi_1(x) < a, \ldots, \varphi_{k-1}(x) < a$, $\varphi_k(x) \geq a$, $\chi_k(x) = 0$ otherwise. $\chi_k(x)$ is an $\mathfrak{A}_n$-measurable function. Hence, for

$k \leq n$

$$\int \varphi_n(x)\, \chi_k(x)\, \mu(dx) = \int \varphi_{n-1}(x)\, \chi_k(x)\, \mu(dx) = \cdots =$$
$$= \int \varphi_k(x)\, \chi_k(x)\, \mu(dx) \geq a \int \chi_k(x)\, \mu(dx) .$$

This means that

$$\int \varphi_n(x) \sum_{k=1}^{n} \chi_k(x)\, \mu(dx) \geq a \int \sum_{k=1}^{n} \chi_k(x)\, \mu(dx) . \tag{8}$$

Clearly, $\sum_{k=1}^{n} \chi_k(x)$ is the indicator of the set $\left\{x \colon \sup_{k\leq n} \varphi_k(x) \geq a\right\} = B_n$. It follows from (1) that

$$\mu(B_n) \leq \frac{1}{a} \int_{B_n} \varphi_n(x)\, \mu(dx) \leq \frac{1}{a} \int \varphi_n^+(x)\, \mu(dx) .$$

The inequality (6) is proved. One gets (7) from (6) by a limit passage on n. □

Theorem 1. *If* $\{\varphi_n(x)\}$ *is a martingale for which* (3) *holds, then* (2) *exists a.e.* (mod μ).

Proof. We will say that the sequence $\alpha_1, \alpha_2, \ldots$ intersects the interval $[\beta_1, \beta_2]$ $(\beta_1 < \beta_2)$ infinitely often if we can find $k_1 < k_2 < \cdots$ such that $\alpha_{k_1} \geq \beta_2$, $\alpha_{k_2} \leq \beta_1, \ldots, \alpha_{k_{2n-1}} \geq \beta_2$, $\alpha_{k_{2n}} \leq \beta_1, \ldots$. Let B_{β_1,β_2} be the set of all x for which the sequence $\{\varphi_n(x)\}$ intersects the interval $[\beta_1, \beta_2]$ infinitely often. Moreover, denote by B_- the set of all x for which $\inf_n \varphi_n(x) = -\infty$, and by B_+ the set of all x for which $\sup_n \varphi_n(x) = +\infty$. Then

$$B_+ \cup B_- \cup \left[\bigcup_{\beta_1<\beta_2} B_{\beta_1,\beta_2}\right],$$

where β_1 and β_2 run through all rational numbers, will be the set of x for which (2) does not exist. It follows from (7) that

$$\mu(B_+) = \lim_{a \uparrow +\infty} \mu\left(\left\{\colon \sup_k \varphi_k(x) \geq a\right\}\right) = 0 .$$

Considering the sequence $-\varphi_n(x)$ we find in exactly the same way that $\mu(B_-) = 0$. Hence, to prove the theorem it suffices to show that for $\beta_1 < \beta_2$

$$\mu(B_{\beta_1,\beta_2}) = 0 . \tag{9}$$

Define a sequence of functions $k_n(x)$ as follows: $k_0(x) = 0$; $k_1(x) = j$ if $\varphi_l(x) < \beta_2$ for $l < j$ and $\varphi_j(x) \geq \beta_2$, $k_1(x) = \infty$ if $\varphi_l(x) < \beta_2$ for all $l > 0$; $k_2(x) = j$ if $k_1(x) < +\infty$ and for $k_1(x) \leq i < j$, $\varphi_i(x) > \beta_1$ and $\varphi_j(x) \leq \beta_1$, otherwise $k_2(x) = +\infty$; if $k_{2n}(x)$ is defined, then $k_{2n+1}(x) = j$ if $k_{2n}(x) < \infty$ and for $k_{2n}(x) \leq i < j$, $\varphi_i(x) < \beta_2$ and

$\varphi_j(x) \geq \beta_2$, if $k_{2n}(x) = +\infty$ or $\varphi_i(x) < \beta_2$ for $i > k_{2n}(x)$, then $k_{2n+1}(x) = +\infty$; if $k_{2n+1}(x)$ is defined, then $k_{2n+2}(x) = j$ if $k_{2n+1}(x) < +\infty$ and for $k_{2n+1}(x) \leq i < j$, $\varphi_i(x) > \beta_1$ and $\varphi_j(x) \leq \beta_1$, if $k_{2n+1}(x) = +\infty$ or $\varphi_i(x) > \beta_1$ for $i > k_{2n+1}(x)$, then $k_{2n+2}(x) = +\infty$. Assume, in addition, that $\chi_n(x) = 1$ if for some i $k_{2i}(x) \leq n < k_{2i+1}(x)$; $\chi_n(x) = -1$ if for some i $k_{2i-1}(x) \leq n < k_{2i}(x)$. It is easy to see that for the determination of $\chi_n(x)$ it is sufficient to know $\varphi_1(x), \ldots, \varphi_n(x)$; moreover, $k_n(x)$ is a Borel function of $\varphi_1(x), \ldots, \varphi_n(x)$. Hence, $\chi_n(x)$ is an $\mathfrak{A}_n$-measurable function. Set

$$\hat{\varphi}_n(x) = \varphi_1(x) + \sum_{k=1}^{n-1} [\varphi_{k+1}(x) - \varphi_k(x)] \, \chi_k(x) \, .$$

We will show that $\hat{\varphi}_n(x)$ is also a martingale. Let $\hat{\mathfrak{A}}_n$ be the σ-algebra generated by the functions $\hat{\varphi}_1(x), \ldots, \hat{\varphi}_n(x)$. Since $\hat{\varphi}_k(x)$ is measurable w.r.t. $\mathfrak{A}_k$, $\hat{\mathfrak{A}}_n \subset \mathfrak{A}_n$. If $\psi(x)$ is bounded and $\hat{\mathfrak{A}}_n$-measurable, then it is also $\mathfrak{A}_n$-measurable. Thus,

$$\int \hat{\varphi}_{n+1}(x) \, \psi(x) \, \mu(dx) = \int [\hat{\varphi}_n(x) + (\varphi_{n+1}(x) - \varphi_n(x)) \, \chi_n(x)] \, \psi(x) \, \mu(dx) = \\ = \int \hat{\varphi}_n(x) \, \psi(x) \, \mu(dx) \, ,$$

since

$$\int [\varphi_{n+1}(x) - \varphi_n(x)] \, \chi_n(x) \, \psi(x) \, \mu(dx) = 0$$

because of the $\mathfrak{A}_n$-measurability and boundedness of $\chi_n(x) \, \psi(x)$. This implies that $\hat{\varphi}_n(x)$ is a martingale. Since $\hat{\varphi}_1(x) = \varphi_1(x)$,

$$\int \hat{\varphi}_n(x) \, \mu(dx) = \int \hat{\varphi}_1(x) \, \mu(dx) = \int \varphi_1(x) \, \mu(dx) \, .$$

Moreover, for $n \leq k_1(x)$, $\hat{\varphi}_n(x) = \varphi_n(x)$. For $k_1(x) < n \leq k_2(x)$, $\hat{\varphi}_n(x)$ $- \hat{\varphi}_{k_1(x)}(x) = \varphi_{k_1(x)}(x) - \varphi_n(x)$ or $\hat{\varphi}_n(x) = 2\,\varphi_{k_1(x)}(x) - \varphi_n(x) \geq 2\,\beta_2 - \varphi_n(x)$. Therefore, $\hat{\varphi}_{k_2(x)}(x) \geq 2\,\beta_2 - \beta_1 > \beta_1 \geq \varphi_{k_2(x)}(x)$ and when $\hat{\varphi}_n(x) \geq \varphi_n(x)$, but $k_3(x) \leq n \leq k_4(x)$, we have $\hat{\varphi}_n(x) - \hat{\varphi}_{k_3(x)}(x) = \varphi_{k_3(x)}(x) - \varphi_n(x)$ and $\hat{\varphi}_n(x) \geq 2\,\varphi_{k_3(x)}(x) - \varphi_n(x) \geq 2\,\beta_2 - \varphi_n(x)$. Analogously, we find that

$$\hat{\varphi}_n(x) \geq \min [\varphi_n(x), 2\,\beta_2 - \varphi_n(x)] \, . \tag{10}$$

Let $\varphi_n(x) = \varphi_n^+(x) - \varphi_n^-(x)$ and $\hat{\varphi}_n(x) = \hat{\varphi}_n^+(x) - \hat{\varphi}_n^-(x)$, where $\varphi_n^+(x) \times \times \varphi_n^-(x) = \hat{\varphi}_n^+(x) \, \hat{\varphi}_n^-(x) = 0$, $\varphi_n^+(x) \geq 0$, $\varphi_n^-(x) \geq 0$, $\hat{\varphi}_n^+(x) \geq 0$, and $\hat{\varphi}_n^-(x) \geq 0$. Then from (10) we obtain

$$\hat{\varphi}_n^-(x) \leq \max [\varphi_n^-(x), \varphi_n^+(x) - 2\,\beta_2) \leq |\varphi_n(x)| + 2\,|\beta_2| \, .$$

Thus,

$$\int \hat{\varphi}_n^+(x) \, \mu(dx) = \int \hat{\varphi}_n(x) \, \mu(dx) + \int \hat{\varphi}_n^-(x) \, \mu(dx) \leq \\ \leq \int [\varphi_n(x) + |\varphi_n(x)| + 2\,|\beta_2|] \, \mu(dx) \leq \\ \leq 2 \int \varphi_n^+(x) \, \mu(dx) + 2\,|\beta_2| \, .$$

This implies that

$$\mu\left(\left\{x: \sup_n \hat{\varphi}_n(x) = +\infty\right\}\right) = \lim_{a\to\infty} \mu\left(\left\{x: \sup_n \hat{\varphi}_n(x) \geq a\right\}\right) \leq$$

$$\leq \lim_{a\to\infty} \frac{1}{a} \sup_n \int \hat{\varphi}_n^+(x)\, \mu(dx) = 0 .$$

Now note that

$$B_{\beta_1,\beta_2} \subset \left\{x: \sup_n \hat{\varphi}_n(x) = +\infty\right\}. \tag{11}$$

Indeed, for $x \in B_{\beta_1,\beta_2}$, $k_n(x) < \infty$ for all n. But for $k_n(x) < +\infty$

$$\hat{\varphi}_{k_n(x)}(x) - \hat{\varphi}_{k_{n-1}(x)}(x) \geq \beta_2 - \beta_1 , \qquad \hat{\varphi}_{k_1(x)} \geq \beta_2 ,$$

so that

$$\hat{\varphi}_{k_n(x)}(x) \geq \beta_2 + (n-1)\,[\beta_2 - \beta_1] .$$

Hence, for $x \in B_{\beta_1,\beta_2}$ $\sup_n \hat{\varphi}_n(x) = +\infty$. We have established (11). Then $\mu(B_{\beta_1,\beta_2}) = 0$. □

Corollary 1. *Any nonnegative martingale* $\{\varphi_n(x)\}$ *possesses a limit a.e.* (*w.r.t.* μ); *if*

$$\varphi(x) = \lim_{n\to\infty} \varphi_n(x) \qquad (\mathrm{mod}\ \mu) ,$$

then

$$\int \varphi(x)\, \mu(dx) \leq \int \varphi_1(x)\, \mu(dx) . \tag{12}$$

In fact, for a nonnegative martingale $\int |\varphi_n(x)| \mu(dx) = \int \varphi_n(x)\, \mu(dx) = \int \varphi_1(x)\, \mu(dx)$ and the condition (3) is always fulfilled. (12) follows from Fatou's lemma.

Corollary 2. *If* $\varphi_n(x)$ *is a nonnegative semi-martingale and*

$$\sup_n \int \varphi_n(x)\, \mu(dx) < \infty ,$$

then the limit

$$\varphi(x) = \lim_{n\to\infty} \varphi_n(x) \qquad (\mathrm{mod}\ \mu)$$

exists and

$$\int \varphi(x)\, \mu(dx) \leq \sup_n \int \varphi_n(x)\, \mu(dx) . \tag{13}$$

If equality holds in (13), *then for any* $\mathfrak{A}_n$*-measurable, bounded nonnegative function* $\psi(x)$

$$\int \varphi(x)\, \psi(x)\, \mu(dx) \geq \int \varphi_n(x)\, \psi(x)\, \mu(dx) . \tag{14}$$

To prove (14) we note that on the basis of Fatou's lemma

$$\int \varphi(x)\, \psi(x)\, \mu(dx) \leq \lim_{m\to\infty} \int \varphi_m(x)\, \psi(x)\, \mu(dx) . \tag{15}$$

Assume that $\alpha > \psi(x)$. Then also

$$\int \varphi(x)\,[\alpha - \psi(x)]\,\mu(dx) \leq \lim_{m\to\infty} \int \varphi_m(x)\,[\alpha - \psi(x)]\,\mu(dx)\,. \tag{16}$$

If strict inequality held in either (15) or (16), then we would have

$$\int \varphi(x)\,\mu(dx) < \lim_{m\to\infty} \int \varphi_m(x)\,\mu(dx)\,,$$

which contradicts the assumption that equality holds in (13). For martingales we have equality in (14) under the same assumptions.

Remark 1. If $\{\varphi_n(x)\}$ is a martingale and $\lambda(\tau)$ is a nonnegative convex function defined on $(-\infty, \infty)$ for which

$$\int \lambda(\varphi_n(x))\,\mu(dx) < \infty\,,$$

for all n, then $\{\lambda(\varphi_n(x))\}$ is a semi-martingale. For any nonnegative $\mathfrak{A}_n$-measurable bounded function $\psi(x)$

$$\begin{aligned}\int \lambda(\varphi_{n+1}(x))\,\psi(x)\,\mu(dx) &= \int \left[\int \lambda(\varphi_{n+1}(y))\,\mu(dy, \mathfrak{A}_n/x)\right] \psi(x)\,\mu(dx) \geq \\ &\geq \int \lambda \left(\int \varphi_{n+1}(y)\,\mu(dy, \mathfrak{A}_n/x\right) \psi(x)\,\mu(dx) = \\ &= \int \lambda(\varphi_n(x))\,\psi(x)\,\mu(dx)\,.\end{aligned}$$

(we have used Jensen's inequality here and the remark following Lemma 1). One can establish in exactly the same way that $\{\lambda(\varphi_n(x))\}$ is also a semi-martingale if $\lambda(\tau)$ is nondecreasing and $\{\varphi_n(x)\}$ is a semi-martingale.

Remark 2. In the proof of Theorem 1 we also showed that

$$\begin{aligned}\mu\left(\{x\colon k_n(x) \leq m\}\right) &= \mu\left(\left\{x\colon \sup_{n\leq m} \hat{\varphi}_n(x) > \beta_2 + (n-1)(\beta_2 - \beta_1)\right\}\right) \leq \\ &\leq \frac{2\int \varphi_m^+(x)\,\mu(dx) + 2\,|\beta_2|}{\beta_2 + (n-1)(\beta_2 - \beta_1)}\,.\end{aligned}$$

But the set $\{x\colon k_n(x) \leq m\}$ coincides with the set of all x for which the sequence $\varphi_1(x), \ldots, \varphi_m(x)$ intersects the interval $[\beta_1, \beta_2]$ no fewer than n times, i.e., for each n one can determine $k_1 < k_2 < \cdots < k_n \leq m$ such that

$$\varphi_{k_1}(x) \geq \beta_2,\quad \varphi_{k_2}(x) \leq \beta_1,\quad \varphi_{k_3}(x) \geq \beta_2\,, \ldots \quad \text{etc.}$$

§ 15. General Conditions for Absolute Continuity

Let $(X, \mathfrak{B})$ be a measurable Hilbert space on which two normalized measures μ^1 and μ^2 are defined. In this section we consider some general conditions for the absolute continuity and singularity of such measures.

If $\mu^2 \ll \mu^1$, than for any finite-dimensional subspace L we have $\mu_L^2 \ll \mu_L^1$, where μ_L^k is the projection of μ^k onto L. On finite-dimensional subspaces the question of absolute continuity can be resolved rather easily. We will assume that the measures μ^1 and μ^2 are such that for all finite-dimensional subspaces the relation $\mu_L^2 \ll \mu_L^1$ holds. We will investigate the absolute continuity of μ^2 w.r.t. μ^1 under this assumption.

Let $\varrho_L(x)$ be the density of μ_L^2 w.r.t. μ_L^1: for all Borel sets $A \subset \mathfrak{B}_L$

$$\mu_L^2(A) = \int_A \varrho_L(x)\, \mu_L^1(dx)\,. \tag{1}$$

(1) is equivalent to the following: for all $A \subset \mathfrak{B}^L$ ($\mathfrak{B}^L$ is the σ-algebra of cylinder sets with bases in L)

$$\mu^2(A) = \int_A \varrho_L(P_L\, x)\, \mu^1(dx) \tag{2}$$

which means that for any $\mathfrak{B}^L$-measurable nonnegative function $\psi(x)$

$$\int \psi(x)\, \mu^2(dx) = \int \varrho_L(P_L\, x)\, \psi(x)\, \mu^1(dx) \tag{3}$$

holds. Introduce a sequence of finite-dimensional subspaces L_n for which $L_n \subset L_{n+1}$ and $\bigcup L_n$ is dense in X. Put

$$\varrho_n(x) = \varrho_{L_n}(P_{L_n}(x))\,.$$

If the function $\psi(x)$ is $\mathfrak{B}^{L_n}$-measurable, then it is also $\mathfrak{B}^{L_{n+1}}$-measurable so that from (3) it follows that

$$\int \varrho_{n+1}(x)\, \psi(x)\, \mu^1(dx) = \int \psi(x)\, \mu^2(dx) = \int \varrho_n(x)\, \psi\,(x)\, \mu^1(dx)\,. \tag{4}$$

Since $\varrho_n(x)$ is $\mathfrak{B}^{L_n}$-measurable, the σ-albegra $\mathfrak{A}_n$, generated by the functions $\rho_1(x), \ldots, \varrho_n(x)$ is contained in the σ-algebra $\mathfrak{B}^{L_n}$. Hence, $\{\varrho_n(x)\}$ forms a martingale on the space with measure $(X, \mathfrak{B}, \mu^1)$. This martingale is nonnegative and from Theorem 1, § 14 the limit

$$\varrho(x) = \lim_{n\to\infty} \varrho_n(x)\,, \tag{5}$$

exists a.e. (w.r.t. μ^1) and

$$\int \varrho(x)\, \mu^1(dx) \le 1$$

since $\int \varrho_n(x)\, \mu^1(dx) = \int \mu^2(dx) = 1$ for all n.

Theorem 1. *For the absolute continuity of μ^2 w. r. t. μ^1 it is necessary and sufficient that the function $\varrho(x)$, defined at* (5), *satisfy the condition*

$$\int \varrho(x)\, \mu^1(dx) = 1\,. \tag{6}$$

In this case

$$\varrho(x) = \frac{d\mu^2}{d\mu^1}(x) \qquad (\mathrm{mod}\, \mu^1) \tag{7}$$

Proof. Let $\mu^2 \ll \mu^1$ and $\pi(x) = \frac{d\mu^2}{d\mu^1}(x)$. Denote by $\mu_n^k(\cdot/x)$ the conditional measure for μ^k w.r.t. the σ-algebra $\mathfrak{B}^{L_n}$. Then for a $\mathfrak{B}^{L_n}$-measurable, nonnegative, bounded function $\psi(x)$, we have

$$\int \psi(x)\, \mu^2(dx) = \int \psi(x)\, \pi(x)\, \mu^1(dx) = \int \psi(x)\, [\int \pi(y)\, \mu_n^1(dy/x)]\, \mu^1(dx)$$

comparing this equality with (3) for $L = L_n$, we obtain

$$\varrho_n(x) = \int \pi(y)\, \mu_n^1(dy/x) \ .$$

We now prove that $\varrho_n(x)$ is uniformly integrable (w.r.t. μ^1) in n. To do this we must show that for any $\varepsilon > 0$ there exists a α such that for all n

$$\int_{\{x:\varrho_n(x)>\alpha\}} \varrho_n(x)\, \mu^1(dx) < \varepsilon \,. \tag{8}$$

Let $\lambda_\alpha(\tau) = 0$ for $\alpha > \tau$ and $\lambda_\alpha(\tau) = \tau - \alpha$ for $\alpha \leq \tau$. Then (8) is equivalent to the following:

$$\int \lambda_\alpha(\varrho_n(x))\, \mu^1(dx) + \alpha\, \mu^1\,(\{x : \varrho_n(x) > \alpha\}) < \varepsilon\,. \tag{9}$$

Since

$$\lambda_\alpha(\varrho_n(x)) = \lambda_\alpha\,(\int \pi(y)\, \mu_n^1(dy/x)) \leq \int \lambda_\alpha(\pi(y))\, \mu_n^1(dy/x)\,,$$

we have

$$\int \lambda_\alpha(\varrho_n(x))\, \mu^1(dx) \leq \int \lambda_\alpha(\pi(y))\, \mu^1(dy)\,.$$

Moreover,

$$\alpha\mu^1(\{x : \varrho_n(x) > \alpha\}) \leq 2 \int \lambda_{\alpha/2}(\varrho_n(x))\, \mu^1(dx) \leq 2 \int \lambda_{\alpha/2}(\pi(x))\, \mu^1(dx)\,.$$

Hence, (9) and then also (8) will hold if α is chosen so that

$$\int \lambda_{\alpha/2}(\pi(x))\, \mu^1(dx) < \frac{\varepsilon}{3}\,.$$

Since $\varrho_n(x)$ is uniformly integrable (w.r.t. μ) in n, it follows that we can go to the limit under the integral sign in the equality

$$\int \varrho_n(x)\, \mu^1(dx) = 1\,.$$

This proves (6). Exactly as for $A \in \mathfrak{B}^{L_n}$, we can proceed to the limit under the integral for $n \to \infty$ in the relation

$$\mu^2(A) = \lim_{n\to\infty} \int \chi_A(x)\, \varrho_n(x)\, \mu^1(dx) = \int \chi_A(x)\, \varrho(x)\, \mu^1(dx)\,,$$

so that (7) is also proved.

Now assume that (6) holds. Then, on the basis of Corollary 2 of Theorem 1 in the preceding section, we have for any $\mathfrak{B}^{L_n}$-measurable, nonnegative bounded function $\psi(x)$

$$\int \varrho(x)\, \psi(x)\, \mu^1(dx) = \int \varrho_n(x)\, \psi(x)\, \mu^1(dx) = \int \psi(x)\, \mu^2(dx)\,.$$

Hence, for any $\mathfrak{B}^{L_n}$-measurable set A

$$\int_A \varrho(x)\,\mu^1(dx) = \int_A \mu^2(dx) = \mu^2(A)\,. \tag{10}$$

Since the class of all sets for which (10) holds contains all of the $\mathfrak{B}^{L_n}$ and is monotone, it also contains the minimal σ-algebra containing all $\mathfrak{B}^{L_n}$, i.e., it contains the σ-algebra $\mathfrak{B}$. From the validity of (10) for all $A \in \mathfrak{B}$ it follows that $\mu^2 \ll \mu^1$ and (7) also hold. □

Corollary. *For the absolute continuity of μ^2 w.r.t. μ^1 it is sufficient that for all n $\mu^2_{L_n} \ll \mu^1_{L_n}$ and that the sequence of functions $\varrho_n(x)$ be uniformly integrable* (w.r.t. μ^1), *i.e., for each $\varepsilon > 0$ there exist a δ such that for all A for which $\mu^1(A) < \delta$*

$$\int_A \varrho_n(x)\,\mu^1(dx) < \varepsilon\,.$$

for all n. Indeed, uniform integrability implies (7). In particular, $\varrho_n(x)$ is uniformly integrable if for some continuous function $\psi(x)$ defined on $[0, \infty)$ and satisfying $\psi(\tau) \to +\infty$ for $\tau \to +\infty$

$$\sup_n \int \varrho_n(x)\,\psi(\varrho_n(x))\,\mu^1(dx) < \infty\,.$$

Another extreme case is when $\varrho(x) = 0 \pmod{\mu^1}$.

Theorem 2. *The equality $\varrho(x) = 0 \pmod{\mu^1}$ is necessary and sufficient for the singularity of μ^2 and μ^1.*

Proof. It $A \in \mathfrak{B}^{L_n}$, then for $n < m$

$$\mu^2(A) = \int_A \varrho_n(x)\,\mu^1(dx)\,.$$

Letting $n \to \infty$ here, we get from Fatou's lemma

$$\mu^2(A) \geq \int_A \varrho(x)\,\mu^1(dx)\,. \tag{11}$$

With the help of a limit passage, (11) can be extended to all $A \in \mathfrak{B}$. Now let $\mu^2 \perp \mu^1$. Then we can find a set A for which $\mu^2(A) = 0$ and $\mu^1(X - A) = 0$. Substituting this A into (11), we find that

$$\int_A \varrho(x)\,\mu^1(dx) = 0\,.$$

Since $\mu^1(X - A) = 0$, it follows from the preceding equality that $\varrho(x) = 0 \pmod{\mu^1}$. The necessity of the condition in the theorem is proved. Now let $\varrho(x) = 0 \pmod{\mu^1}$. We will show that $\mu^1 \perp \mu^2$. Assume this is not the case. We can represent μ^2 in the form

$$\mu^2 = \nu' + (1 - \alpha)\,\nu''\,,$$

where $\nu' \ll \mu^1$, $\nu'' \perp \mu^1$ and $0 < \alpha \leq 1$. Then, setting

$$\varrho_n'(x) = \frac{d\nu_{L_n}'}{d\mu_{L_n}^1}(P_{L_n}x),$$

we obtain by Theorem 1

$$\lim_{n\to\infty} \varrho_n'(x) = \frac{d\nu'}{d\mu^1}(x) \qquad (\text{mod}\ \mu^1).$$

But $\alpha\, \varrho_n'(x) \leq \varrho_n(x)$ so that

$$\lim_{n\to\infty} \varrho_n'(x) \leq \frac{1}{\alpha} \lim_{n\to\infty} \varrho_n(x) = 0,$$

i.e., $\frac{d\nu'}{d\mu^1}(x) = 0$ (mod μ^1), which contradicts the assumption on the absolute continuity of ν' w.r.t. μ^1. Hence $\mu^1 \perp \mu^2$. □

From the last two theorems follows

Theorem 3. *The function $\varrho(x)$, defined by (3), is the density of the absolutely continuous component of the measure μ^2 w.r.t. the measure μ^1, i.e., (7) holds in all cases.*

Proof. Let $\mu^2 = \alpha\, \nu' + (1-\alpha)\, \nu''$, where $\nu' \ll \mu^1$, $\nu'' \perp \mu^1$ and $0 \leq \alpha < 1$. Let ν_{L_n}' and ν_{L_n}'' be the projections of the measures ν' and ν'' on L_n and

$$\varrho_n'(x) = \frac{d\nu_{L_n}'}{d\mu_{L_n}^1}(P_{L_n}x) \quad \text{and} \quad \varrho_n''(x) = \frac{d\nu_{L_n}''}{d\mu_{L_n}^1}(P_{L_n}x).$$

Then

$$\varrho_n(x) = \alpha\, \varrho_n'(x) + (1-\alpha)\, \varrho_n''(x).$$

By Theorem 1

$$\lim_{n\to\infty} \varrho_n'(x) = \frac{d\nu'}{d\mu^1}(x) \qquad (\text{mod}\ \mu^1),$$

and Theorem 2 asserts that

$$\lim_{n\to\infty} \varrho_n''(x) = 0 \qquad (\text{mod}\ \mu^1).$$

Hence,

$$\lim_{n\to\infty} \varrho_n(x) = \alpha\, \frac{d\nu'}{d\mu^1}(x) = \frac{d\mu^2}{d\mu^1}(x) \qquad (\text{mod}\ \mu^1). \quad \square$$

Up to now we have considered only the case in which $\mu_{L_n}^2 \ll \mu_{L_n}^1$ for all n. Assume this no longer holds. We will then show how one can construct a decomposition of the measure μ^2 into two components: $\mu^2 = \nu' + \nu''$ such that $\nu_{L_n}'' \ll \mu_{L_n}^1$ for all n and the measure ν' is such that

$$\lim_{n\to\infty} \int \frac{d\nu_{L_n}'}{d\mu_{L_n}^1}(x)\, \mu_{L_n}^1(dx) = 0. \tag{12}$$

Let $A_n \in \mathfrak{B}_{L_n}$ be a set for which $\mu^1_{L_n}(A_n) = 0$ and $\mu^2_{L_n}(B - A_n)$ is absolutely continuous w.r.t. $\mu^1_{L_n}(B)$ on $\mathfrak{B}_{L_n}$. Further let $\widetilde{A}_n$ be a cylinder set from the σ-algebra $\mathfrak{B}_{L_n}$ with base A_n. Put $\widetilde{A} = \bigcup_{n=1}^{\infty} \widetilde{A}_n$, $\nu''(B) = \mu^2(B - \widetilde{A})$ and $\nu'(B) = \mu^2(B \cap \widetilde{A})$. Then

$$\nu''_{L_n}(A) \ll \mu^2_{L_n}(A - A_n)$$

and consequently, $\nu''_{L_n} \ll \mu^1_{L_n}$ (according to the construction of the set A_n). Further, since $\mu^1_{L_n}(P^{-1}_{L_k} A_k \cap L_n) = 0$ for $k \le n$ and $\frac{d\nu'_{L_n}}{d\mu^1_{L_n}}$ is different from zero only on $P_{L_n}\widetilde{A}$, we get

$$\int \frac{d\nu'_{L_n}}{d\mu^1_{L_n}}(x)\, \mu^1_{L_n}(dx) \le \nu'_{L_n}\left(P_{L_n}\left[\widetilde{A} - \bigcup_{k=1}^{n} \widetilde{A}_k\right]\right),$$

since for any pair of measures μ and ν

$$\int_A \frac{d\nu}{d\mu}(x)\, \mu(dx) \le \nu(A)\,.$$

Denote by C_n a cylinder set in $\mathfrak{B}^{L_n}$ with base $P_{L_n}\left[\widetilde{A} - \bigcup_{k=1}^{n} \widetilde{A}_k\right]$. Obviously, $C_n \supset C_{n+1}$ and $\lim_{n\to\infty} \nu'(C_n) = \nu'\left(\bigcap_n C_n\right)$. Since C_n has an empty intersection with $\widetilde{A}_n$, $\bigcap_n C_n$ also has an empty intersection with each $\widetilde{A}_k$ and hence also with $\bigcup_k \widetilde{A}_k = A$. Hence,

$$\nu'\left(\bigcap_n C_n\right) = \nu'\left(\widetilde{A} \cap \left[\bigcap_n C_n\right]\right) = 0\,.$$

Since

$$\int \frac{d\nu'}{d\mu^1_{L_n}}(x)\, \mu^1_{L_n}(dx) \le \nu'(C_n)\,,$$

we have thereby proved (12).

The measure ν' represents "the sum of the finite-dimensional singularities of the measure μ^2 w.r.t. μ^1". It is obviously singular w.r.t. μ^1. For the measure ν'', the assertions of Theorems 1, 2 and 3 hold. It follows from the construction of the measure ν' that for $x \notin \left[\bigcup_n C_n\right] \cup \widetilde{A}$

$$\lim_{n\to\infty} \frac{d\nu'_{L_n}}{d\mu^1_{L_n}}(P_{L_n} x) = 0\,. \tag{13}$$

Since $\nu'\left(\bigcap_n C_n\right) = 0$ and $\mu^1(\widetilde{A}) = 0$, (13) is satisfied a.e. (w.r.t. μ^1). Hence,

$$\lim_{n\to\infty} \frac{d\mu^2_{L_n}}{d\mu^1_{L_n}}(x) = \lim_{n\to\infty} \frac{d\nu''_{L_n}}{d\mu^1_{L_n}}(x) = \frac{d\nu''}{d\mu^1}(x) = \frac{d\mu^2}{d\mu^1}(x) \qquad (\text{mod}\ \mu^1)\,.$$

We have proved

Theorem 4. *Let μ^1 and μ^2 be arbitrary normalized measures on $(X, \mathfrak{B})$ and L_n an increasing sequence of finite-dimensional subspaces for which $\bigcup L_n$ is dense in X. Then*

$$\frac{d\mu^2}{d\mu^1}(x) = \lim_{n\to\infty} \frac{d\mu^2_{L_n}}{d\mu^1_{L_n}}(x) \qquad (\mathrm{mod}\ \mu^1) \tag{14}$$

and $\mu^2 \ll \mu^1$ iff

$$\int \frac{d\mu^2}{d\mu^1}(x)\, \mu^1(dx) = 1\,.$$

To calculate the density of one measure w.r.t. another we use the properties of densities under certain transformations of measures.

Theorem 5. *Assume the measurable Hilbert space $(X, \mathfrak{B})$ is the Cartesian product of two separable, measurable Hilbert spaces $(X_1, \mathfrak{B}_1) \times \times (X_2, \mathfrak{B}_2)$ on each of which are given two measures:* ν^1 and μ^1 on $(X_1, \mathfrak{B}_1)$ *and ν^2 and μ^2 on $(X_2, \mathfrak{B}_2)$. Let $\mu = \mu^1 \times \mu^2$ and $\nu = \nu^1 \times \nu^2$ be measures on $(X, \mathfrak{B})$. Then $\nu \ll \mu$ iff $\nu^1 \ll \mu^1$ and $\nu^2 \ll \mu^2$ and we have*

$$\frac{d\nu}{d\mu}(x) = \frac{d\nu^1}{d\mu^1}(P_1 x)\, \frac{d\nu^2}{d\mu^2}(P_2 x)\,, \tag{15}$$

where $P_1 x = x^1 \in X_1$ and $P_2 x = x^2 \in X_2$ if $x = (x^1, x^2)$.

Proof. If $\nu \ll \mu$, then $\nu(A_1 \times X_2) = \nu^1(A_1) = 0$ provided that $\mu(A_1 \times X_2) = \mu^1(A_1) = 0$, $A_1 \in \mathfrak{B}_1$. Hence, $\nu^1 \ll \mu^1$. Likewise, $\nu^2 \ll \mu^2$. The necessity part of the theorem is proved. Now assume $\nu^1 \ll \mu^1$ and $\nu^2 \ll \mu^2$. Denote by $\mathfrak{B}^0$ the algebra of sets from $\mathfrak{B}$ representable as sums of the form

$$\bigcup_{k=1}^{n} A^1_k \times A^2_k\,, \tag{16}$$

where $A^1_k \in \mathfrak{B}_1$ and $A^2_k \in \mathfrak{B}_2$. Each set from $\mathfrak{B}^0$ can be written as the sum of non-overlapping sets contained in the sum (16). Let

$$\varrho(x) = \frac{d\nu^1}{d\mu^1}(P_1 x)\, \frac{d\nu^2}{d\mu^2}(P_2 x)\,.$$

Since

$$\int\limits_{A^1 \times A^2} \varrho(x)\, \mu(dx) = \int\limits_{A^1} \frac{d\nu^1}{d\mu^1}(x^1)\, \mu^1(dx^1) \int\limits_{A^2} \frac{d\nu^2}{d\mu^2}(x^2)\, \mu^2(dx^2) = \nu^1(A^1)\, \nu^2(A^2) = $$

$$= \nu(A^1 \times A^2)\,,$$

one has for all $A \in \mathfrak{B}^0$,

$$\int\limits_A \varrho(x)\, \mu(dx) = \nu(A)\,. \tag{17}$$

Relation (17) holds on a monotone class of sets containing $\mathfrak{B}^0$; but then it is satisfied for all $A \in \mathfrak{B}$. From this it follows that $\nu \ll \mu$, whence (15). □

Another transformation which finds frequent application is that of a measure under a mapping of the corresponding spaces. We have already shown in § 13 that under a map f of the space $(X, \mathfrak{B})$ into $(Y, \mathfrak{C})$ the properties of absolute continuity or singularity are preserved for the pair of measures μ^1 and μ^2 on $(X, \mathfrak{B})$ and ν^1, ν^2 on $(Y, \mathfrak{C})$, where

$$\nu^k(C) = \mu^k(f^{-1}(C)), \quad C \in \mathfrak{C}. \tag{18}$$

We will show how to calculate the density of the transformed measures w.r.t. one another when the density of the original measures is known.

Theorem 6. *If μ^1 and μ^2 are defined on $(X, \mathfrak{B})$ with $\mu^2 \ll \mu^1$ and ν^1 and ν^2 are measures on $(Y, \mathfrak{C})$ defined by* (18), *where f is a measurable transformation of $(X, \mathfrak{B})$ into $(Y, \mathfrak{C})$, then*

$$\frac{d\nu^2}{d\nu^1}(y) = \int \frac{d\mu^2}{d\mu^1}(x)\, \mu^1(dx, \mathfrak{B}_1/f^{-1}(y)), \tag{19}$$

where $\mu^1(\cdot, \mathfrak{B}_1/z)$ is the conditional measure for μ^1 w.r.t. the σ-algebra $\mathfrak{B}_1$ generated by sets of the form $f^{-1}(C)$, $C \in \mathfrak{C}$.

Proof. We first show that we can substitute $f^{-1}(y)$ into the expression $\mu^1(A, \mathfrak{B}_1/z)$, $A \in \mathfrak{B}$, for any f. Let $A_y = \{z : f(z) = y\}$, $A_y \in \mathfrak{B}_1$ and for any $\mathfrak{B}_1$-measurable set A^1 assume either $A_y \subset A^1$ or $A_y = X - A^1$. This means that each $\mathfrak{B}_1$-measurable function is constant on A_y and because of the $\mathfrak{B}_1$-measurability in z of $\mu^1(A, \mathfrak{B}_1/z)$, the expression $\mu^1(A, \mathfrak{B}_1/f^{-1}(y))$ does not depend on the choice of the inverse image $f^{-1}(y)$ of the point y.

Let $\psi(y)$ be a bounded $\mathfrak{C}$-measurable function on $(Y, \mathfrak{C})$. Then the function $\psi(f(x)) = \varphi(x)$ will be a $\mathfrak{B}_1$-measurable bounded function on $(X, \mathfrak{B})$. Consequently,

$$\int \psi(y)\, \nu^2(dy) = \int \psi(f(x))\, \mu^2(dx) = \int \psi(f(x)) \frac{d\mu^2}{d\mu^1}(x)\, \mu^1(dx).$$

Using (10) § 13 we have

$$\int \psi(f(x)) \frac{d\mu^2}{d\mu^1}(x)\, \mu^1(dx) = \int \psi(f(z)) \int \frac{d\mu^2}{d\mu^1}(x)\, \mu^1(dx, \mathfrak{B}_1/z)\, \mu^1(dz).$$

Assume

$$\int \frac{d\mu^2}{d\mu^1}(x)\, \mu^1(dx, \mathfrak{B}_1/f^{-1}(y)) = \varrho'(y). \tag{20}$$

(the possibility of substituting $f^{-1}(y)$ into this expression has already been discussed). Then

$$\int \psi(f(z))\, \varrho'(f(z))\, \mu^1(dz) = \int \psi(y)\, \varrho'(y)\, \nu^1(dy),$$

so that

$$\int \psi(y)\, \nu^2(dy) = \int \psi(y)\, \varrho'(y)\, \nu^1(dy)\,.$$

From this and (20) we obtain the proof of the theorem. □

With the help of Theorems 5 and 6 we can find an expression for the density of convoluted measures. Let two measures μ^1 and μ^2 be given on $(X, \mathfrak{B})$. Denote by $A - x$ the set $\{y : y + x \in A\}$. It is easy to verify that the function $\mu^1(A - x)$ is $\mathfrak{B}$-measurable. The measure $\mu^1 * \mu^2$, defined by means of

$$\mu^1 * \mu^2(A) = \int \mu^2 (A - x)\, \mu^1(dx) \tag{21}$$

is called the *convolution* of the measures μ^1 and μ^2. The operation of convolution is commutative. To show this we will derive another expression for $\mu^1 * \mu^2$. Denote by $\mu^1 \times \mu^2$ the product of these measures on $(X, \mathfrak{B}) \times (X, \mathfrak{B})$. Elements of this product space will be written as $(x^1; x^2)$. The set $S^A = \{(x^1; x^2) : x^1 + x^2 \in A\}$ is $\mathfrak{B} \times \mathfrak{B}$-measurable for $\mathfrak{B}$-measurable A. $S^A_{x^1}$ is the "section of this set w.r.t. the first coordinate", i.e., $\{x^2 : (x^1; x^2) \in S^A\}$, and has the form $A - x^1$. By the definition of the product measure

$$\mu^1 \times \mu^2(S^A) = \int \mu^2(S^A_{x^1})\, \mu^1(dx^1) = \int \mu^2 (A - x^1)\, \mu^1(dx^1)\,. \tag{22}$$

In the same way we can define

$$\mu^1 \times \mu^2(S^A) = \int \mu^1(S^A_{x^2})\, \mu^2(dx) = \int \mu^1 (A - x^2)\, \mu^2(dx^2)\,. \tag{23}$$

From this we also obtain commutativity of the convolution since $\mu^2 \times \mu^1(A)$ stands on the left in (23). We remark finally, that $\mu^1 * \mu^2$ can be obtained under a transformation of the space $(X, \mathfrak{B}) \times (X, \mathfrak{B})$ into $(X, \mathfrak{B})$ by means of the function $f(x^1; x^2) = x^1 + x^2$.

Corollary. *Assume yet another pair of measures ν^1 and ν^2 is given on $(X, \mathfrak{B})$, whereby $\nu^1 \ll \mu^1$ and $\nu^2 \ll \mu^2$. Then $\nu^1 \times \nu^2 \ll \mu^1 \times \mu^2$ and*

$$\frac{d\nu^1 \times \nu^2}{d\mu^1 \times \mu^2}(x^1; x^2) = \frac{d\nu^1}{d\mu^1}(x^1)\, \frac{d\nu^2}{d\mu^2}(x^2)\,.$$

On the basis of Theorem 6 we now conclude that $\nu^1 * \nu^2 \ll \mu^1 * \mu^2$ and

$$\frac{d\nu^1 * \nu^2}{d\mu^1 * \mu^2}(x) = \int \frac{d\nu^1}{d\mu^1}(x^1)\, \frac{d\nu^2}{d\mu^2}(x^2)\, \mu^1 \times \mu^2 \left(dx^1 \times dx^2, (\mathfrak{B} \times \mathfrak{B})^+ / \sigma^{-1}(x)\right), \tag{24}$$

where $\sigma(x^1; x^2) = x^1 + x^2$ and $(\mathfrak{B} \times \mathfrak{B})^+$ is the σ-algebra generated by the sets $\sigma^{-1}(A)$, $A \in \mathfrak{B}$.

§ 16. Absolute Continuity of Product Measures

Assume the measures μ and ν on a Hilbert space $(X, \mathfrak{B})$ satisfy the following conditions: It is possible to find a complete sequence of orthogonal subspaces X_k such that for each n

$$\mu_{L_n} = \prod_{k=1}^{n} \mu_{X_k}, \qquad \nu_{L_n} = \prod_{k=1}^{n} \nu_{X_k}, \tag{1}$$

where $L_n = X_1 + \cdots + X_n$ and μ_{L_n}, μ_{X_k}, ν_{L_n} and ν_{X_k} are respectively the projections of the measures μ and ν on the subspaces L_n and X_k and the subspace L_n can be identified with the Cartesian product of the subspaces X_k, $k = 1, \ldots, n$. Formulae (1) say that the measures μ_{L_n} and ν_{L_n} are the products of the measures μ_{X_k} and ν_{X_k}, resp. Measures satisfying the above condition are called *product measures*. They can be written as an infinite product of the form

$$\mu = \prod_{k=1}^{\infty} \mu_{X_k} \tag{2}$$

(this notation refers merely to the method of obtaining the projections μ_{L_n}; it cannot be interpreted in the usual sense since X cannot be considered as $\prod_{k=1}^{\infty} X_k$. However, we can assume that $X \subset \prod_{k=1}^{\infty} X_k$ and then (2) can be understood in the usual measure-theoretic way).

The subspaces X_k can be either finite- or infinite-dimensional. For brevity we will write $\mu_{X_k} = \mu_k$ and $\nu_{X_k} = \nu_k$. The goal of this section is to find conditions for the absolute continuity and singularity of the measures μ and ν expressed in terms of the measures μ_k and ν_k. We first find conditions for the absolute continuity of ν w.r.t. μ. If $\nu \ll \mu$, then for all k $\nu_k \ll \mu_k$.
Let

$$\varrho_k(x) = \frac{d\nu_k}{d\mu_k}(x), \qquad x \in X_k.$$

Then from Theorem 5 § 15

$$\frac{d\nu_{L_n}}{d\mu_{L_n}}(x) = \prod_{k=1}^{n} \varrho_k(P_k x), \qquad x \in L_n,$$

where P_k is projection on X_k. As established in § 15, the sequence $\frac{d\nu_{L_n}}{d\mu_{L_n}}(P_{L_n} x)$ is a martingale on the space with measure $(X, \mathfrak{B}, \mu)$ (it's easy to see that finite-dimensionality was not used in the proof of this in § 15). Hence, on the basis of the corollary to Theorem 1 § 14, the limit

$$\lim_{n\to\infty} \prod_{k=1}^{n} \varrho_k(P_k x) = \prod_{k=1}^{\infty} \varrho_k(P_k x), \tag{3}$$

exists almost everywhere (w.r.t. μ). This limit can also equal zero, so that the infinite product on the right either converges, or diverges to zero. Let us study the convergence of this product. Let B_k be the set of all $x \in X_k$ for which $\varrho_k(x) = 0$.

Lemma 1. *If*

$$\sum_{k=1}^{\infty} \mu_k(B_k) = +\infty\,, \qquad \text{then} \qquad \prod_{k=1}^{\infty} \varrho_k(P_k\, x) = 0 \qquad (\mathrm{mod}\ \mu)\,.$$

Proof. Let B_k' be a cylinder set with base B_k. It is clear that for $x \in \bigcup_{k=1}^{\infty} B_k'$, at least one of the factors $\varrho_k(P_k\, x)$ is equal to zero so that the infinite product in (3) is equal to zero. We show that $\mu\left(\bigcup_{k=1}^{\infty} B_k'\right) = 1$. We have

$$\mu\left(\bigcup_{k=1}^{n} B_k'\right) = \mu_{L_n}\left(P_{L_n} \bigcup_{k=1}^{n} B_k'\right) = 1 - \mu_{L_n}\left(L_n - P_{L_n} \bigcup_{k=1}^{n} B_k'\right) =$$

$$= 1 - \mu_{L_n}\left(\bigcap_{k=1}^{n} [L_n - P_{L_n} B_k']\right) =$$

$$= 1 - \mu_{L_n}\left(\prod_{k=1}^{n} (X_k - B_k)\right) = 1 - \prod_{k=1}^{n} \mu_k\,(X_k - B_k) =$$

$$= 1 - \prod_{k=1}^{n} (1 - \mu_k(B_k))\,.$$

Hence,

$$\mu\left(\bigcup_{k=1}^{\infty} B_k'\right) = \lim_{n\to\infty}\left[1 - \prod_{k=1}^{n} (1 - \mu(B_k))\right] = 1\,,$$

since

$$\prod_{k=1}^{n} (1 - \mu(B_k)) \leq \exp\left\{-\sum_{k=1}^{n} \mu(B_k)\right\} \to 0$$

for $n \to \infty$ by hypothesis. □

Corollary. *If* $\nu \ll \mu$, *then* $\sum_{k=1}^{\infty} \mu_k(B_k) < \infty$.

Assume now that $B' = \bigcup_k B_k'$ and $\sum \mu_k(B_k) < \infty$. Introduce a new measure $\bar{\mu}$ defined by

$$\bar{\mu}(A) = \mu\,(A - B')/\mu\,(X - B')\,. \tag{4}$$

If $\sum \mu_k(B_k) < \infty$, then it follows from the proof of Lemma 1 that

$$\mu\,(X - B') = \prod_{k=1}^{\infty} (1 - \mu_k(B_k)) > 0\,.$$

Hence, (4) makes sense. It's easy to see that for any sequence of measurable sets $A_k \subset X_k$ we have, denoting by A_k' a cylinder set with base A_k,

$$\bar{\mu}_{L_n}(A_1' \cap A_2' \cap \cdots \cap A_n') = \frac{\mu_{L_n}((A_1' - B_1') \cap \cdots \cap (A_n' - B_n'))}{\mu_{L_n}(L_n - P_{L_n} B')} =$$

$$= \frac{\prod_{k=1}^{n} \mu_k(A_k - B_k)}{\prod_{k=1}^{n} (1 - \mu_k(B_k))} = \prod_{k=1}^{n} \bar{\mu}_{X_k}(A_k) .$$

Thus, $\bar{\mu}$ is also a product measure. Setting $\bar{\mu}_k = \bar{\mu}_{X_k}$, we note that $\nu_k \ll \bar{\mu}_k$ and

$$\frac{d\nu_k}{d\bar{\mu}_k}(x) = \varrho_k(x)\,[1 - \mu_k(B_k)]^{-1} \qquad (\mathrm{mod}\ \bar{\mu}) .$$

Since $\prod_{k=1}^{\infty} (1 - \mu_k(B_k))^{-1}$ is convergent,

$$\frac{d\nu}{d\bar{\mu}}(x) = \frac{d\nu}{d\mu}(x) \prod_{k=1}^{\infty} (1 - \mu_k(B_k))^{-1} \qquad (\mathrm{mod}\ \bar{\mu}) ,$$

so that $\nu \ll \mu$ iff $\nu \ll \bar{\mu}$.

Lemma 2. *Either*

$$\prod_{k=1}^{\infty} \frac{d\nu_k}{d\bar{\mu}_k}(P_k x) = 0 \qquad (\mathrm{mod}\ \bar{\mu})$$

or

$$\prod_{k=1}^{\infty} \frac{d\nu_k}{d\bar{\mu}}(P_k x) > 0 \qquad (\mathrm{mod}\ \bar{\mu}) .$$

Proof. Let $\mathfrak{C}^n$ be the σ-algebra of Borel sets generated by cylinder sets with bases in X_k, $k = n, n+1, \ldots$, and $\mathfrak{C}_{n-1}$ the σ-algebra generated by cylinder sets with bases in $X_1, \ldots, X_{n-1}$. If A^n is an arbitrary set from $\mathfrak{C}^n$ and A_{n-1} a set in $\mathfrak{C}_{n-1}$, then

$$\bar{\mu}(A_{n-1} \cap A_n) = \bar{\mu}(A_{n-1})\,\bar{\mu}(A_n) . \tag{5}$$

Formula (5) is easily verified for the cylinder sets A_{n-1} and A^n with bases of the form $C_1 \times \cdots \times C_{n-1}$ or $C^n \times \cdots \times C^{n+N}$, where the C_k are Borel sets in X_k. Then (5) is extended by continuity to the σ-algebras $\mathfrak{C}_{n-1}$ and $\mathfrak{C}^n$. Let $\mathfrak{C}^\infty = \bigcap_n \mathfrak{C}^n$. $\mathfrak{C}^\infty$ is also a σ-algebra. If A^∞ is measurable w.r.t. $\mathfrak{C}^\infty$, then for all n and any $\mathfrak{C}^n$-measurable set A

$$\bar{\mu}(A \cap A^\infty) = \bar{\mu}(A)\,\bar{\mu}(A^\infty) . \tag{6}$$

This means that (6) also holds for A measurable w.r.t. the σ-algebra generated by the σ-algebras $\mathfrak{C}_n$, i.e., for all $\mathfrak{B}$-measurable A. In particular, we can take $A = A^\infty$. Then

$$(\bar{\mu}(A^\infty))^2 = \bar{\mu}(A^\infty) . \tag{7}$$

From (7) it follows that for any $A^\infty \in \mathfrak{C}^\infty$, $\bar{\mu}(A^\infty)$ equals either 0 or 1. Since for all k

$$\frac{d\nu_k}{d\bar{\mu}_k}(P_k x) > 0 \qquad (\mathrm{mod}\ \bar{\mu}) ,$$

we find that

$$\prod_{k=1}^{\infty} \frac{d\nu_k}{d\bar{\mu}_k}(P_k x) > 0$$

iff $\prod_{k=n}^{\infty} \frac{d\nu_k}{d\bar{\mu}_k}(P_k x) > 0$ for all n. The function $\prod_{k=n}^{\infty} \frac{d\nu_k}{d\bar{\mu}_k}(P_k x)$ is $\mathfrak{C}^n$-measurable. Hence, $\left\{x : \prod_{k=1}^{\infty} \frac{d\nu_k}{d\bar{\mu}_k}(P_k x) > 0\right\} \in \mathfrak{C}^n$ for all n. Thus,

$$\left\{x : \prod_{k=1}^{\infty} \frac{d\nu_k}{d\bar{\mu}_k}(P_k x) > 0\right\} \in \mathfrak{C}^\infty .$$

We have shown that

$$\bar{\mu}\left(\left\{x : \prod_{k=1}^{\infty} \frac{d\nu_k}{d\bar{\mu}_k}(P_k x) > 0\right\}\right)$$

is either zero or one. □

Remark. In the course of the proof of the lemma it was established that for product measures sets in $\mathfrak{C}^\infty$ have measure 0 or 1. This property is called *Kolmogorov's zero-one law.*

We can now formulate a necessary and sufficient condition for the absolute continuity of the measure ν w.r.t. the measure μ.

Theorem 1. *In order for ν to be absolutely continuous w.r.t. μ it is necessary and sufficient that the following two conditions hold:*

a) *for all k $\nu_k \ll \mu_k$;*

b) *for some $\alpha \in (0, 1)$ the infinite product*

$$\prod_{k=1}^{\infty} \left(\int \left[\frac{d\nu_k}{d\mu_k}(P_k x) \right]^\alpha \mu(dx) \right) . \tag{8}$$

converges (to a value different from zero).

Proof. The necessity of a) is obvious. If $\nu \ll \mu$, then

$$\frac{d\nu}{d\mu}(x) = \prod_{k=1}^{\infty} \frac{d\nu_k}{d\mu_k}(P_k x) ,$$

so that

$$\left(\frac{d\nu}{d\mu}(x)\right)^\alpha = \lim_{n\to\infty} \prod_{k=1}^{n} \left[\frac{d\nu_k}{d\mu_k}(P_k x)\right]^\alpha . \tag{9}$$

The function $\psi_n(x) = \prod_{k=1}^{n} \left[\frac{d\nu_k}{d\mu_k}(P_k x)\right]^\alpha$ is uniformly integrable w.r.t. n since the integrals

$$\int [\psi_n(x)]^{1/\alpha} \mu(dx) = 1 \qquad \left(\frac{1}{\alpha} > 1\right)$$

are uniformly bounded. Thus, (9) can be integrated taking the limit operation outside the integral:

$$\int \left(\frac{d\bar{\nu}}{d\mu}(x)\right)^\alpha \mu(dx) = \lim_{n\to\infty} \int \prod_{k=1}^{n} \left[\frac{d\nu_k}{d\mu_k}(P_k x)\right]^\alpha \mu(dx) =$$

$$= \lim_{n\to\infty} \prod_{k=1}^{n} \int \left[\frac{d\nu_k}{d\mu_k}(P_k x)\right]^\alpha \mu(dx) .$$

This establishes the necessity of b). We now turn to the sufficiency of the hypotheses. From Lemmas 1 and 2 it follows that when a) is satisfied the following alternatives hold: either $\nu \ll \mu$ or $\nu \perp \mu$. The latter will be the case if

$$\sum_{k=1}^{\infty} \mu_k(B_k) = +\infty \qquad \text{or} \qquad \prod_{k=1}^{\infty} \frac{d\nu_k}{d\mu_k}(P_k x) = 0 \qquad (\text{mod } \mu) .$$

Hence, to prove sufficiency we must show that if $\nu \perp \mu$ and a) holds, then the infinite product (8) diverges to zero. But in that case

$$\lim_{n\to\infty} \prod_{k=1}^{n} \frac{d\nu_k}{d\mu_k}(P_k x) = 0 \qquad (\text{mod } \mu)$$

so that

$$\lim_{n\to\infty} \prod_{k=1}^{n} \left(\frac{d\nu_k}{d\mu_k}(P_k x)\right)^\alpha = 0 . \tag{10}$$

Again using the uniform integrability of the function $\prod_{k=1}^{n} \left(\frac{d\nu_k}{d\mu_k}(P_k x)\right)^\alpha$, we find that (10) can be integrated taking the limit outside the integral. Finally then,

$$\lim_{n\to\infty} \int \prod_{k=1}^{n} \left(\frac{d\nu_k}{d\mu_k}(P_k x)\right)^\alpha \mu(dx) = \lim_{n\to\infty} \prod_{k=1}^{n} \int \left(\frac{d\nu_k}{d\mu_k}(P_k x)\right)^\alpha \mu(dx)$$

and the theorem is completely proved. □

Remark. Fundamental in the proof of Theorem 1 (sufficiency) was the fact that for the measures ν and μ the alternatives $\nu \ll \mu$ or $\nu \perp \mu$ held. It's easy to see that when these alternatives hold for the product measures μ^1 and μ^2, then an assertion analogous to Theorem 1 can be obtained. Let us formulate the assertion.

If μ^1 and μ^2 are product measures on $(X, \mathfrak{B})$ for which the alternatives $\mu^2 \ll \mu^1$ or $\mu^1 \perp \mu^2$ hold, then $\mu^2 \ll \mu^1$ iff the following conditions are satisfied:

a) *for some increasing sequence of finite-dimensional subspaces L_n for which $\bigcup L_n$ is dense in X, $\mu^2_{L_n} \ll \mu^1_{L_n}$;*

b) *for some $0 < \alpha < 1$, the limit*

$$\lim_{n\to\infty} \int \left(\frac{d\mu^2_{Ln}}{d\mu^1_{Ln}} (P_{L_n} x)\right)^\alpha \mu^1(dx)$$

exists and is different from zero (it is necessary that these conditions be satisfied for an arbitrary sequence of subspaces L_n and a number α, sufficient that there exist at least one sequence L_n and number α satisfying a) *and* b)).

As an example of the application of Theorem 1 we consider conditions for the absolute continuity of two Gaussian measures with different mean values. To this end we need to define the expression $(b, A^{-1/2} x)$, where A is the correlation operator of a Gaussian measure μ with mean 0. Let $\{e_k\}$ be the sequence of eigenvectors of the operator A and λ_k the corresponding eigenvalues. We then take

$$(b, A^{-1/2} x) = \lim_{n\to\infty} \sum_{k=1}^{n} \frac{(x, e_k)(b, e_k)}{\sqrt{\lambda_k}}, \tag{11}$$

with the limit in the sense of convergence w.r.t. μ. This limit even exists in mean square (w.r.t. μ) since

$$\int \left(\sum_{k=n}^{n+m} \frac{(x, e_k)(b, e_k)}{\sqrt{\lambda_k}}\right)^2 \mu(dx) = \int \sum_{k,j=n}^{n+m} \frac{(x, e_k)(b, e_k)(x, e_j)(b, e_j)}{\sqrt{\lambda_k \lambda_j}} \mu(dx) =$$

$$= \int \sum_{k=n}^{n+m} \frac{(x, e_k)^2}{\lambda_k} (b, e_k)^2 \mu(dx) = \sum_{k=n}^{n+m} \frac{(b, e_k)^2}{\lambda_k} (A e_k, e_k) = \sum_{k=n}^{n+m} (b, e_k)^2 .$$

The last expression tends to zero for $n \to \infty$, uniformly in n, since

$$\sum_{k=1}^{\infty} (b, e_k)^2 = (b, b) < \infty .$$

It is not difficult to verify that the sequence of functions $\sum_{k=1}^{n} \frac{(x, e_k)(b, e_k)}{\sqrt{\lambda_k}}$ forms a martingale, so that the limit exists μ — a.e.

Theorem 2. *Let μ and ν be Gaussian measures on $(X, \mathfrak{B})$ whose characteristic functionals have the forms*

$$\int e^{i(z,x)} \mu(dx) = \exp\left\{-\frac{1}{2}(A z, z)\right\}$$

and

$$\int e^{i(z,x)}\,\nu(dx) = \exp\left\{i(a, z) - \frac{1}{2}(A\,z, z)\right\}.$$

If a vector $b \in X$ exists for which $a = A^{1/2}\,b$, then $\nu \sim \mu$ and

$$\frac{d\nu}{d\mu}(x) = \exp\left\{(b, A^{-1/2}\,x) - \frac{1}{2}(b, b)\right\}, \tag{12}$$

otherwise $\nu \perp \mu$.

Proof. The measures μ and ν turn out to be product measures if we take as X_k, $k = 0, \ldots$, the one-dimensional eigen-subspaces of the operator A corresponding for $k > 0$ to the non-zero eigenvalues and as X_0 the eigen-subspace corresponding to the eigenvalue 0. This fact was established in § 5. Denote by μ_k and ν_k, $k = 0, 1, \ldots$, the projections of the measures μ and ν on the subspace X_k. Since μ_0 is Gaussian in X_0 with mean zero and correlation operator 0, it is concentrated at the point 0. $\nu \ll \mu$ only if ν is also concentrated at 0. This will happen only when $P_{X_0}\,a = 0$. Hence, $\nu \ll \mu$ only if a can be decomposed in terms of the eigenvectors $\{e_k\}$ of the operator A corresponding to non-zero eigenvalues. Let $(a, e_k) = \alpha_k$. Then, as was established in § 5, the measures μ_k and ν_k ($k > 0$) will be absolutely continuous w.r.t. Lebesgue measure on X_k with densities having the values

$$\frac{1}{\sqrt{2\pi\lambda_k}}\exp\{-\tau^2/2\lambda_k\} \qquad \text{and} \qquad \frac{1}{\sqrt{2\pi\lambda_k}}\exp\left\{-\frac{(\tau-\alpha_k)^2}{2\lambda_k}\right\}$$

at the point $x = \tau\,e_k$. Thus, for $k > 0$, $\frac{d\nu_k}{d\mu_k}(x)$ always exists and

$$\frac{d\nu_k}{d\mu_k}(\tau\,e_k) = \exp\left\{\frac{\tau\,\alpha_k}{\lambda_k} - \frac{1}{2\lambda_k}\alpha_k^2\right\}.$$

For $0 < \gamma < 1$

$$\int\left(\frac{d\nu_k}{d\mu_k}(P_k\,x)\right)^{\gamma}\mu(dx) = \int\left(\frac{d\nu_k}{d\mu_k}(x)\right)^{\gamma}\mu(dx) =$$

$$= \int_{-\infty}^{\infty}\left(\exp\left\{\frac{\tau\,\alpha_k}{\lambda_k} - \frac{1}{2\lambda_k}\alpha_k^2\right\}\right)^{\gamma}\frac{1}{\sqrt{2\pi\lambda_k}}\exp\left\{-\frac{\tau^2}{2\lambda_k}\right\}d\tau =$$

$$= \frac{1}{\sqrt{2\pi\lambda_k}}\exp\left\{-\frac{\gamma}{2\lambda_k}\alpha_k^2\right\}\int_{-\infty}^{\infty}\exp\left\{-\frac{\tau^2}{2\lambda_k} - 2\gamma\,\alpha_k\,\tau\right\}d\tau =$$

$$= \exp\left\{-\frac{\gamma(1-\gamma)}{2\lambda_k}\alpha_k^2\right\} = \exp\left\{-\frac{\gamma(1-\gamma)}{2\lambda_k}(a, e_k)^2\right\}.$$

On the basis of Theorem 1, we can show that $\nu \ll \mu$ iff $P_{X_0}\,a = 0$ and

$$\prod_{k=1}^{\infty}\exp\left\{-\frac{\gamma(1-\gamma)}{2\lambda_k}(a, e_k)^2\right\} = \exp\left\{-\frac{\gamma(1-\gamma)}{2}\sum\frac{(a, e_k)^2}{\lambda_k}\right\} > 0,$$

i.e., $P_{X_0} a = 0$ and $\sum_{k=1}^{\infty} \frac{(a, e_k)^2}{\lambda_k} < \infty$. Assume that

$$b = \sum_{k=1}^{\infty} \frac{(a, e_k)}{\sqrt{\lambda_k}} e_k .$$

Then

$$A^{1/2} b = \sum_{k=1}^{\infty} \frac{(a, e_k)}{\sqrt{\lambda_k}} A^{1/2} e_k = \sum_{k=1}^{\infty} \frac{(a, e_k)}{\sqrt{\lambda_k}} \sqrt{\lambda_k}\, e_k = a .$$

Hence, $\nu \ll \mu$ iff there exists a $b \in X$ for which $A^{1/2} b = a$. Then

$$\frac{d\nu}{d\mu}(x) = \prod_{k=1}^{\infty} \exp\left\{\frac{(x, e_k)(a, e_k)}{\lambda_k} - \frac{1}{2\lambda_k}\alpha_k^2\right\} =$$

$$= \exp\left\{-\frac{1}{2}(b, b) + \lim_{n\to\infty} \sum_{1}^{n} \frac{(x, e_k)(a, e_k)}{\lambda_k}\right\} =$$

$$= \exp\left\{(b, A^{-1/2} x) - \frac{1}{2}(b, b)\right\}.$$

If ν is not absolutely continuous w.r.t. μ, then $\nu \perp \mu$, as was proved in Theorem 1. Finally, when $\nu \ll \mu$, it follows from (12) that $\frac{d\nu}{d\mu}(x) > 0$ (mod μ), so that $\mu \ll \nu$. Hence, $\nu \sim \mu$ and the theorem is completely proved. □

§ 17. Absolute Continuity of Gaussian Measures

In the previous section we considered conditions for the absolute continuity of Gaussian measures with identical correlation operators. Here we will consider Gaussian measures with not only different means but also different correlation operators. Denote by $\mu(a, A; \cdot)$ a Gaussian measure with mean a and correlation operator A. We will see that the investigation of the absolute continuity of Gaussian measures can always be reduced to that of the absolute continuity of Gaussian measures having either the same means or the same correlation operators.

Theorem 1. *If* $\mu(a_1, A_1; \cdot) \ll \mu(a, A; \cdot)$, *then* $\mu(a_1, A_1; \cdot) \ll \mu(a_1, A; \cdot)$ *and* $\mu(a_1, A; \cdot) \ll \mu(a, A; \cdot)$.

Proof. Using the fact that under one-to-one transformations of a space absolutely continuous measures go into absolutely continuous measures, we can write: $\mu(a_1 - a, A_1; \cdot) \ll \mu(0, A; \cdot)$, $\mu(a - a_1, A_1; \cdot) \ll \mu(0, A; \cdot)$, where we applied the transformation $f(x) = x - a$ in the first case and $f(x) = -x$ in the second. From the corollary to Theorem 6,

§ 15 we have

$$\mu(a_1 - a, A_1; \cdot) * \mu(a - a_1, A_1; \cdot) \ll \mu(0, A; \cdot) * \mu(0, A; \cdot) . \quad (1)$$

Employing (22) § 15, we easily discover that for any two measures μ^1 and μ^2

$$\int e^{i(z,x)} \mu^1 * \mu^2(dx) = \int \int e^{i(z,x+y)} \mu^1(dx) \mu^2(dy) = \\ = \int e^{i(z,x)} \mu^1(dx) \int e^{i(z,y)} \mu^2(dy) . \quad (2)$$

From (2) in shortened notation

$$\mu(a', A') * \mu(a'', A'') = \mu(a' + a'', A' + A'') \quad (3)$$

which means that (1) is equivalent to the relation

$$\mu(0, 2A_1) \ll \mu(0, 2A) .$$

Applying the transformation $f(x) = \frac{1}{\sqrt{2}} x$, we obtain $\mu(0, A_1) \ll \mu(0, A)$, so that also $\mu(a_1, A_1) \ll \mu(a_1, A)$. Since $\mu(a_1, A_1) \ll \mu(a, A)$, $\mu(a_1, A) \perp \mu(a, A)$ is not possible since we would then also have $\mu(a_1, A_1) \perp \mu(a, A)$. But from Theorem 2, § 16, the measures $\mu(a_1, A)$ and $\mu(a, A)$ must be either equivalent or orthogonal. Hence, $\mu(a_1, A; \cdot) \sim \mu(a, A; \cdot)$. □

We will now consider the case in which $a = a_1 = 0$ (all other cases can be reduced to this one by a translation of the space). Note that if $\mu(0, A_2) \ll \mu(0, A_1)$, then the subspaces H_i concide, where H_i is the closed range of the operator A_i. If there exists a $z \in X$ for which $(A_k z, z) = 0$ and $(A_j z, z) > 0$, then

$$\mu(0, A_k; \{x: (z, x) = 0\}) = 1 ; \qquad \mu(0, A_j; \{x: (z, x) = 0 \}) = 0$$

so that $\mu(0, A_k) \perp \mu(0, A_j)$. Hence, without loss of generality, we can assume that $H_1 = H_2 = X$, i.e., that both measures are nondegenerate.

Let L_n be an arbitrary increasing sequence of finite-dimensional subspaces with $\bigcup L_n$ dense in X, P_n the projection operator onto L_n and $A_k^{(n)} = P_n A_k$ the correlation operator considered in L_n. Since A is a nonsingular operator, $A_k^{(n)}$ is an invertible symmetric operator for all n. Set $\mu_n^k = \mu_{L_n}(0, A_k)$. The characteristic function of the measure μ_n^k in the finite-dimensional space L_n has the form

$$\int e^{i(z,x)} \mu_n^k(dx) = \exp\left\{-\frac{1}{2} (A_k^{(n)} z, z)\right\}, \qquad z \in L_n .$$

From the results of § 5 it follows that the measure μ_n^k has a density w.r.t. Lebesgue measure in L_n of the form

$$(2\pi)^{-n/2} [\det A_k^{(n)}]^{-1/2} \exp\left\{-\frac{1}{2} (A_k^{(n)-1} x, x)\right\}, \qquad x \in L_n , \quad (4)$$

where $\det A_k^{(n)}$ is the determinant of the matrix of the operator $A_k^{(n)}$ in an orthonormalized basis in L_n. Consequently, for all n $\mu_n^2 \ll \mu_n^1$ and

$$\frac{d\mu_n^2}{d\mu_n^1}(x) = [\det A_1^{(n)} A_2^{(n)}]^{-1} \exp\left\{\frac{1}{2}([A_1^{(n)^{-1}} - A_2^{(n)^{-1}}]\, x, x)\right\}, \quad x \in L_n .$$

Hence,

$$\log \frac{d\mu_n^2}{d\mu_n^1}(P_n x) = (B_n x, x) + \delta_n ,$$

where

$$B_n = [A_1^{(n)^{-1}} - A_2^{(n)^{-1}}]\, P_n \quad \text{and} \quad \delta_n \doteq \frac{1}{2} \log [\det A_1^{(n)} / \det A_2^{(n)}] . \tag{5}$$

If $\mu(0, A_2) \ll \mu(0, A_1)$, then $(B_n x, x) + \delta_n$ converges in $\mu(0, A_1)$ to some limit, possibily equal to $-\infty$, and on a set of positive measure this limit is finite. Using this fact, we prove the important

Theorem 2. *The following alternatives hold: either* $\mu(0, A_2) \sim \mu(0, A_1)$ *or* $\mu(0, A_2) \perp \mu(0, A_1)$.

Proof. We investigate the behavior of $(B_n x, x) + \delta_n$, where B_n is a sequence of singular symmetric operators. Consider in L_n the scalar product $\langle x, y\rangle = (x, A_1^{(n)^{-1}} y)$. The quadratic form $(B_n x, x)$ can be reduced to the major axis in the scalar product $\langle .,. \rangle$. Then

$$(B_n x, x) = \sum_1^n \sigma_k^{(n)} \langle x, f_k^{(n)}\rangle^2 , \tag{6}$$

where $f_1^{(n)}, \ldots, f_n^{(n)} \in L_n$ and $\langle f_k^{(n)}, f_j^{(n)}\rangle = \delta_{kj}$. Note that for $z \in L_n$

$$\int \exp\left\{i \sum_1^n \tau_k \langle x, f_k^{(n)}\rangle\right\} \mu_n^1(dx) =$$

$$\int \exp\left\{i \left(x, \sum_1^n \tau_k A_1^{(n)^{-1}} f_k^{(n)}\right)\right\} \mu_n^1(dx) =$$

$$\exp\left\{-\frac{1}{2}\left(A_1^{(n)} \sum_1^n \tau_k A_1^{(n)^{-1}} f_k^{(n)}, \sum_1^n \tau_k A_1^{(n)^{-1}} f_k^{(n)}\right)\right\} =$$

$$\exp\left\{-\frac{1}{2} \sum_1^n \tau_k^2\right\} =$$

$$(2\pi)^{-n/2} \int \cdots \int \exp\left\{i \sum_1^n \tau_k \xi_k\right\} \exp\left\{-\frac{1}{2} \sum_1^n \xi_k^2\right\} d\xi_1 \cdots d\xi_n . \tag{7}$$

From this relation it follows that under the transformation of L_n into R^n (n-dimensional Euclidean space) defined by the formula

$$f(x) = (\langle x, f_1\rangle, \ldots, \langle x, f_n\rangle) ,$$

the measure $\mu_1^{(n)}(dx)$ goes into the measure $\gamma(B)$ defined on the set $B \in \mathfrak{B}_{R^n}$ by the equality

$$\gamma(B) = \int_B \cdots \int (2\pi)^{-n/2} \exp\left\{-\frac{1}{2} \sum_1^n \xi_k^2\right\} d\xi_1 \cdots d\xi_n$$

(on the right is an n-tuple Lebesgue integral). Thus for any Borel function $\varphi(\xi_1, \ldots, \xi_n)$ on R^n.

$$\int \varphi(\langle x, f_1\rangle, \ldots, \langle x, f_n\rangle)\, \mu_n^2(dx) = $$
$$= \int \cdots \int (2\pi)^{-n/2} \varphi(\xi_1, \ldots, \xi_n)\, e^{-\frac{1}{2}\sum_1^n \xi_k^2}\, d\xi_1 \cdots d\xi_n \tag{8}$$

provided that the integral on the right exists. Using (6) and (7), we obtain

$$\int e^{(B_n x, x)} \mu_n^1(dx) = \prod_1^n \frac{1}{1 - 2\sigma_k^{(n)}},$$

whereby the integral on the left exists if $\sigma_k^{(n)} < \frac{1}{2}$. From the relation

$$\int e^{(B_n x, x) + \delta_n} \mu_n^1(dx) = 1$$

we find that $\sigma_k^{(n)} < \frac{1}{2}$ and

$$\delta_n = \sum_1^n \log(1 - 2\sigma_k^{(n)}) . \tag{9}$$

It is easy to see from (8) and (9) that for $0 < \alpha < 1$

$$\int \left(\frac{d\mu_n^2}{d\mu_n^1}(P_n x)\right)^\alpha \mu(0, A; dx) = e^{\alpha\delta_n} \int e^{\alpha(B_n x, x)} \mu_n^1(dx) =$$
$$= e^{\alpha \sum_1^n \log\left(1 - 2\sigma_k^{(n)}\right)} \prod_1^n \frac{1}{1 - 2\alpha\sigma_k^{(n)}}. \tag{10}$$

Assume that

$$\lim_{n\to\infty} \frac{d\mu_n^2}{d\mu_n^1}(P_n x)$$

is not a.e. equal to 0 $\big(\text{mod } \mu(0, A_1)\big)$ (the existence of this limit was established in § 15). Then, for $\alpha < 1$

$$\lim_{n\to\infty} \int \left(\frac{d\mu_n^2}{d\mu_n^1}(P_n x)\right)^\alpha \mu(0, A_1; dx) = \int \left(\lim_{n\to\infty} \frac{d\mu_n^2}{d\mu_n^1}(P_n x)\right)^\alpha \mu(0, A_1; dx) > 0$$

(the possibility of proceeding to the limit under the integral sign follows from the uniform integrability of the integrand, which in turn follows from the same arguments as in the remark following Theorem 1 § 16). Consequently, in this case the expression

$$\sum_1^n [\alpha \log(1 - 2\sigma_k^{(n)}) - \log(1 - 2\alpha\sigma_k^{(n)})] .$$

is bounded w.r.t. n. Since for $\xi > 0$ and $0 < \alpha < 1$ the equality $\xi^\alpha \le 1 - \alpha + \alpha\xi$ holds, for $\sigma_k^{(n)} < 1$ we have $(1 - 2\sigma_k^{(n)})^\alpha \le 1 - 2\alpha\sigma_k^{(n)} = 1 - \alpha + \alpha(1 - 2\sigma_k^{(n)})$, so that

$$\sum_1^n \alpha \log(1 - 2\sigma_k^{(n)}) - \log(1 - 2\alpha\sigma_k^{(n)}) = \sum_1^n \log \frac{(1 - 2\sigma_k^{(n)})^\alpha}{1 - 2\alpha\sigma_k^{(n)}}, \tag{11}$$

where the latter is a sum of nonpositive terms. This means that all terms in this sum are also bounded so that there exists a $\beta > 0$ such that

$$-\beta \leq \log \frac{(1-2\sigma_k^{(n)})^\alpha}{1-2\alpha\sigma_k^{(n)}}.$$

From this follows the existence of an $\varepsilon > 0$ for which $-1/\varepsilon < 2\sigma_k^{(n)} < 1-\varepsilon$. It's easy to see that the function $-\frac{1}{\xi^2}\log\frac{(1-\xi)^\alpha}{1-\alpha\xi}$ is continuous and positive for $\xi \in [-1/\varepsilon, 1-\varepsilon]$ if extended by continuity at $\xi = 0$. This means that there exist constants β_1 and β_2 for which

$$\beta_1(\sigma_k^{(n)})^2 \leq -\log\frac{(1-2\sigma_k^{(n)})^\alpha}{1-2\alpha\sigma_k^{(n)}} \leq \beta_2(\sigma_k^{(n)})^2.$$

Hence, from the boundedness of the sum (11) there follows the boundedness of $\sum_1^n (\sigma_k^{(n)})^2$ and the existence of an $\varepsilon > 0$ such that

$$2\sigma_k^{(n)} < 1-\varepsilon.$$

We note that (10) holds for all α for which $2\alpha\sigma_k^{(n)} < 1$. It therefore holds if $\alpha(1-\varepsilon) < 1$. In particular,

$$\int \left(\frac{d\mu_n^2}{d\mu_n^1}(P_n x)\right)^{\frac{1}{1-\varepsilon/2}} \mu(0, A_1; dx) = \exp\left\{\sum_1^n \log\frac{(1-2\sigma_k^{(n)})^{\frac{1}{1-\varepsilon/2}}}{1-\frac{2}{1-\varepsilon\sigma_k^{(n)}/2}}\right\} \leq$$

$$\leq \exp\left\{\beta_3 \sum_1^n (\sigma_k^{(n)})^2\right\}, \tag{12}$$

where β_3 is some constant. The corollary to Theorem 1, § 15 then implies that $\mu(0, A_2) \ll \mu(0, A_1)$ $\left(\text{as } \psi(\tau) \text{ we can take } \tau^{\frac{1}{1-\varepsilon/2}-1}\right)$. We have therefore proved the following assertion: if $\mu(0, A_1)$ and $\mu(0, A_2)$ are not mutually singular, then $\mu(0, A_2) \ll \mu(0, A_1)$. Reversing the indices 1 and 2 we find also that $\mu(0, A_1) \ll \mu(0, A_2)$. Hence, if $\mu(0, A_1)$ and $\mu(0, A_2)$ are not mutually singular, they are equivalent. □

We investigate the form of the density when $\mu(0, A_1)$ and $\mu(0, A_2)$ are equivalent. Since in this case

$$\log\frac{d\mu(0, A_2)}{d\mu(0, A_1)}(x)$$

is defined a.e. (w.r.t. $\mu(0, A_1)$), we have

$$\frac{d\mu(0, A_2)}{d\mu(0, A_1)}(x) = \exp\left\{\lim_{n\to\infty} [(B_n x, x) + \delta_n]\right\},$$

where B_n and δ_n are defined at (5). The limit

$$\lim_{n\to\infty} [(B_n x, x) + \delta_n]$$

exists a.e. $(\mathrm{mod}\, \mu(0, A_1))$. We will show that the mean square limit (in μ) also exists. We use the representation (6). It was established in the proof of Theorem 2 that when the measures are equivalent, there exists an $\varepsilon > 0$ such that $-\frac{1}{\varepsilon} < 2\,\sigma_k^{(n)} < 1 - \varepsilon$. Choose α such that $4\alpha|\sigma_k^{(n)}| < 1$ for all k and n. Then, as in (12) we obtain

$$\begin{aligned}\int \exp\{\alpha\,[(B_n x, x) + \delta_n] - \alpha\,[(B_m x, x) + \delta_m]\}\, \mu(0, A_1; dx) \leq \\ \leq \Big(\int \exp\{2\alpha(B_n x, x) + 2\alpha\delta_n\}\, \mu(0, A_1; dx) \times \\ \times \int \exp\{-2\alpha(B_m x, x) - 2\alpha\delta_m\}\, \mu(0, A_1; dx)\Big)^{1/2} \leq \\ \leq \exp\left\{\beta_4 \sum_1^n (\sigma_k^{(n)})^2 + \beta_5 \sum_1^m (\sigma_k^{(m)})^2\right\},\end{aligned}$$

where β_4 and β_5 are some constants. Hence, the set of functions

$$\exp\left\{\frac{\alpha}{2}\,[(B_n x, x) + \delta_n] - \frac{\alpha}{2}\,[(B_m x, x) + \delta_m]\right\}$$

are uniformly integrable and tend to zero for $n \to \infty$, $m \to \infty$. Thus,

$$\lim_{n,m\to\infty} \int \exp\left\{\frac{\alpha}{2}\,[(B_n x, x) + \delta_n] - \frac{\alpha}{2}\,[(B_m x, x) + \delta_m]\right\} \mu(0, A_1; dx) = 1\,. \tag{13}$$

From (13) we have

$$\begin{aligned}\lim_{n,m\to\infty} \int \Big(\exp\left\{\frac{\alpha}{2}\,[(B_n x, x) + \delta_n] - \frac{\alpha}{2}\,[(B_m x, x) + \delta_m]\right\} + \\ + \exp\left\{\frac{\alpha}{2}\,[(B_m x, x) + \delta_m) - \frac{\alpha}{2}\,[(B_n x, x) + \delta_n]\right\} - 2\Big)\, \mu(0, A_1; dx) = 0\,.\end{aligned} \tag{14}$$

Finally, using the inequality

$$e^{\xi} + e^{-\xi} - 2 \geq \xi^2\,,$$

we obtain from (14)

$$\lim_{n,m\to\infty} \int [(B_n x, x) + \delta_n - (B_m x, x) - \delta_m]^2\, \mu(0, A_1; dx) = 0\,. \tag{15}$$

Hence we have proved the existence of the mean square limit in $\mu(0, A_1)$ of the sequence $(B_n x, x) + \delta_n$.

Let L_n be a sequence of subspaces spanned by $e_1, \ldots, e_n$, where $\{e_k\}$ are the eigenvectors of the operator A_1. Then

$$(B_n x, x) + \delta_n = \sum \beta_{kj}^{(n)} \left[\frac{(x, e_k)}{\sqrt{\lambda_k}} \cdot \frac{(x, e_j)}{\sqrt{\lambda_j}} - \delta_{kj}\right] + \eta_n\,,$$

where λ_k is the eigenvalue of A_1 corresponding to the eigenvector e_k and

$$\beta_{kj}^{(n)} = (B_n e_k, e_j)\sqrt{\lambda_k \lambda_j}, \qquad \eta_n = \delta_n + \sum_{k=1}^{n} \beta_{kk}^{(n)}.$$

Starting from the relation

$$\int (x, e_k)(x, e_j)(x, e_l)(x, e_m)\,\mu(0, A_1; dx) =$$

$$= \frac{\partial^4}{\partial\tau_1\,\partial\tau_2\,\partial\tau_3\,\partial\tau_4} \int e^{i(x, \tau_1 e_k + \tau_2 e_j + \tau_3 e_l + \tau_4 e_m)}\,\mu(0, A_1; dx)\Big|_{\tau_1=\tau_2=\tau_3=\tau_4=0} =$$

$$= \frac{\partial^4}{\partial\tau_1\,\partial\tau_2\,\partial\tau_3\,\partial\tau_4} \exp \times$$

$$\times \left\{-\frac{1}{2}(A_1(\tau_1 e_k + \tau_2 e_j + \tau_3 e_l + \tau_4 e_m), \tau_1 e_k + \tau_2 e_j + \tau_3 e_l + \tau_4 e_m)\right\}\Big|_{\tau_1=\tau_2=\tau_3=\tau_4=0}$$

we find that

$$\int (x, e_k)(x, e_j)(x, e_l)(x, e_m)\,\mu(0, A_1; dx)$$

is different from zero only in the cases $k = j$, $l = m$; $k = l$, $j = m$ and $k = m$, $j = l$; when $k = l$ and $j = m$, it equals $\lambda_j \lambda_k$ and when $k = j = l = m$, it equals $3\lambda_k^2$. Hence,

$$\int [(B_n x, x) + \delta_n - (B_m x, x) - \delta_m]^2\,\mu(0, A_1; dx) =$$

$$= 2\sum_{k,j} (\beta_{kj}^{(n)} - \beta_{kj}^{(m)})^2 + (\eta_n - \eta_m)^2.$$

This implies that there exist numbers β_{kj} and η such that

$$\lim_{n\to\infty} \eta_n = \eta \qquad \text{and} \qquad \lim_{n\to\infty} \sum_{k,j} (\beta_{kj}^{(n)} - \beta_{kj})^2 = 0. \tag{16}$$

Let

$$\psi(x) = \eta + \sum_{k,j} \beta_{kj}\left[\frac{(x, e_k)}{\sqrt{\lambda_k}} \cdot \frac{(x, e_j)}{\sqrt{\lambda_j}} - \delta_{kj}\right]. \tag{17}$$

The series in (17) converges since the functions $\left[\frac{(x, e_k)\,(x, e_j)}{\sqrt{\lambda_k}\sqrt{\lambda_j}} - \delta_{kj}\right]$ are orthonormalized w.r.t. the measure $\mu(0, A_1)$ and $\sum \beta_{kj}^2 < \infty$. Since

$$\int [\psi(x) - (B_n x, x) - \delta_n]^2\,\mu(0, A_1; dx) = (\eta - \eta_n)^2 + \sum_{k,j} (\beta_{kj}^{(n)} - \beta_{kj})^2,$$

(16) implies that

$$\psi(x) = \lim\,[(B_n x, x) + \delta_n]$$

in the sense of mean square convergence in $\mu(0, A_1)$. We have proved

Theorem 3. *If* $\mu(0, A_1) \sim \mu(0, A_2)$, *then there exist numbers* η and β_{kj}, $j, k = 1, 2, \ldots$, *satisfying the condition*

$$\sum_{k,j} \beta_{kj}^2 < \infty$$

for which

$$\frac{d\mu(0, A_2)}{d\mu(0, A_1)}(x) = \exp\{\psi(x)\}, \tag{18}$$

where $\psi(x)$ *is defined by* (17).

Remark. The constant η in (17) can be found from the relation

$$\int \exp\{\psi(x)\}\, \mu(0, A_1; dx) = 1\,.$$

In order to explain the connection between the function $\psi(x)$ and the correlation operators A_1 and A_2, we evaluate the integral

$$\int \exp\{i(z, x) + \psi(x)\}\, \mu(0, A_1; dx) = \int \exp\{i(z, x)\}\, \mu(0, A_2; dx) = \\ = \exp\left\{-\frac{1}{2}(A_2 z, z)\right\}.$$

Let B be an operator for which

$$(B e_k, e_j) = \beta_{kj}\,.$$

Since $\beta_{kj} = \beta_{jk}$, this operator is symmetric. Moreover,

$$(B^2 e_k, e_k) = (B e_k, B e_k) = \Big(\sum_j \beta_{kj} e_j, \sum_j \beta_{kj} e_j\Big) = \sum_j (\beta_{kj})^2\,.$$

Thus,

$$\sum_k (B^2 e_k, e_k) = \sum_{k,j} \beta_{kj}^2 < \infty\,.$$

Hence, B^2 is nuclear and B is a completely continuous operator. Let $\{f_k\}$ be a complete orthonormalized system of eigenvectors of the operator B and β_k the corresponding eigenvalues. We will show that $\psi(x)$ can be represented as

$$\psi(x) = \eta + \sum_{k=1}^{\infty} \beta_k\,[(f_k, A^{-1/2} x)^2 - 1]\,, \tag{19}$$

where $(f_k, A^{-1/2} x)$ is determined by means of (11) § 16. Indeed,

$$\sum_{k=1}^{\infty} \beta_k\,[(f_k, A^{-1/2} x)^2 - 1] = \sum_{k=1}^{\infty} \beta_k \left[\left(\sum_{j=1}^{\infty}\left(f_k, \sum_{j=1}^{\infty} \frac{(x, e_j)}{\sqrt{\lambda_j}} e_j\right)\right)^2 - 1\right] = \\ = \sum_{k=1}^{\infty} \beta_k \sum_{j,l=1}^{\infty} (f_k, e_j)\,(f_k, e_l)\left[\frac{(x, e_j)}{\sqrt{\lambda_j}} \cdot \frac{(x, e_l)}{\sqrt{\lambda_l}} - \delta_{jl}\right]$$

since

$$\sum_{j=1}^{\infty} (f_k, e_j)^2 = (f_k, f_k) = 1\,.$$

Moreover,

$$\lim_{n\to\infty} \sum_{k=1}^{n} \beta_k \sum_{j,l=1}^{\infty} (f_k, e_j)(f_k, e_l)\left[\frac{(x, e_k)(x, e_l)}{\sqrt{\lambda_j \lambda_l}} - \delta_{jl}\right] =$$

$$= \lim_{n\to\infty} \sum_{l,j=1}^{\infty} \sum_{k=1}^{n} \beta_k (f_k, e_j)(f_k, e_l)\left[\frac{(x, e_j)(x, e_l)}{\sqrt{\lambda_j \lambda_l}} - 1\right]. \tag{20}$$

Let P_n be the projection operator on the space spanned by $f_1, \ldots, f_n$. Then

$$\sum_{k=1}^{n} \beta_k (f_k, e_j)(f_k, e_l) = \sum_{k=1}^{\infty} (B f_k, e_j)(P_n f_k, e_l) =$$

$$= \sum_{k=1}^{\infty} (f_k, B e_j)(f_k, P_n e_l) = (B e_j, P_n e_l) = (P_n B e_j, e_l),$$

$$\sum_{j,l} [(P_n B e_j, e_l) - (B e_j, e_l)]^2 = \sum_{j,l} ([I - P_n] B e_j, e_l)^2 =$$

$$= \operatorname{tr}([I - P_n] B)^2 = \sum_{k=1}^{\infty} ((I - P_n) B f_k, (I - P_n) B f_k) =$$

$$= \sum_{k=n+1}^{\infty} \beta_k^2 \to 0 \qquad \text{for } n \to \infty .$$

Consequently, the limit (20) equals

$$\sum_{j,l=1}^{\infty} (B e_j, e_l)\left[\frac{(x, e_j)(x, e_l)}{\sqrt{\lambda_j \lambda_l}} - \delta_{jl}\right] = \sum_{j,l} \beta_{jl}\left[\frac{(x, e_j)(x, e_l)}{\sqrt{\lambda_j \lambda_l}} - \delta_{jl}\right].$$

This establishes (19). Note that for all n

$$\int \exp\left\{i \sum_{k=1}^{n} \tau_k (f_k, A_1^{-1/2} x)\right\} \mu(0, A_1; dx) =$$

$$= \lim_{m\to\infty} \int \exp\left\{i \sum_{k=1}^{n} \tau_k \left(f_k, \sum_{1}^{m} \frac{(x, e_j)}{\sqrt{\lambda_j}} e_j\right)\right\} \mu(0, A_1; dx) =$$

$$= \lim_{m\to\infty} \int \exp\left\{i \left(x, \sum_{k=1}^{n} \tau_k \sum_{1}^{m} \frac{(f_k, e_j)}{\sqrt{\lambda_j}} e_j\right)\right\} \mu(0, A_1; dx) =$$

$$= \lim_{m\to\infty} \exp\left\{-\frac{1}{2}\left(A_1 \sum_{k=1}^{n} \tau_k \sum_{1}^{m} \frac{(f_k, e_j)}{\sqrt{\lambda_j}} e_j, \sum_{k=1}^{n} \tau_k \sum_{1}^{m} \frac{(f_k, e_j)}{\sqrt{\lambda_j}} e_j\right)\right\} =$$

$$= \lim_{m\to\infty} \exp\left\{-\frac{1}{2} \sum_{j=1}^{m} \left(\sum_{1}^{n} \tau_k f_k, e_j\right)^2\right\} =$$

$$= \exp\left\{-\frac{1}{2}\left(\sum_{1}^{n} \tau_k f_k, \sum_{1}^{n} \tau_k f_k\right)\right\} = \exp\left\{-\frac{1}{2} \sum \tau_k^2\right\}.$$

Starting with this formula we can obtain the following one in exactly the same way as (8) was derived from (7): for any Borel function $\varphi(\xi_1, \ldots, \xi_n)$ in R^n

$$\int \varphi((f_1, A_1^{-1/2} x), \ldots, (f_n, A_1^{-1/2} x))\, \mu(0, A_1; dx) =$$
$$= \int \cdots \int (2\pi)^{-n/2} \varphi(\xi_1, \ldots, \xi_n)\, e^{-\frac{1}{2}\sum_1^n \xi_k^2}\, d\xi_1 \cdots d\xi_n ,$$

provided the latter integral exists. Using this, we find that

$$\int \exp\left\{\sum_{k=1}^{\infty} \beta_k [(f_k, A_1^{-1/2} x)^2 - 1]\right\} \mu(0, A_1; dx) = \prod_{k=1}^{\infty} \frac{e^{-\beta_k}}{\sqrt{1 - 2\beta_k}}$$

(the integral exists for $2\beta_k < 1$). Thus,

$$\eta = \log \prod_{k=1}^{\infty} \frac{e^{-\beta_k}}{\sqrt{1 - 2\beta_k}} = \sum_{k=1}^{\infty} \left[-\beta_k - \frac{1}{2} \log(1 - 2\beta_k)\right]$$

(the convergence of the series follows from the estimate $\beta_k + \frac{1}{2} \times$ $\times \log(1 - 2\beta_k) = O(\beta_k^2)$ and the fact that $\sum \beta_k^2 < \infty$). Moreover,

$$\int \exp\{i(z, x) + \psi(x)\}\, \mu(0, A_1; dx) =$$
$$= \int \exp\left\{i \sum_{k=1}^{\infty} (A_1^{1/2} z, f_k)(A_1^{1/2} x, f_k) + \right.$$
$$\left. + \sum_{k=1}^{\infty} \beta_k [(A_1^{-1/2} x, f_k)^2 - 1]\right\} \mu(0, A_1; dx) =$$
$$= e^{\eta} \prod_{k=1}^{\infty} \frac{1}{\sqrt{2\pi}} \int e^{i(\xi(A_1^{1/2} z, f_k)) + \beta_k(\xi^2 - 1) - \xi^2/2}\, d\xi =$$
$$= e^{\eta} \prod_{k=1}^{\infty} \frac{1}{\sqrt{1 - 2\beta_k}} \exp\left\{-\beta_k - \frac{1}{2} \frac{(A_1^{1/2} z, f_k)^2}{1 - 2\beta_k}\right\} =$$
$$= \exp\left\{-\frac{1}{2} \sum_{k=1}^{\infty} \frac{(A_1^{1/2} z, f_k)^2}{1 - 2\beta_k}\right\} =$$
$$= \exp\left\{-\frac{1}{2} \sum_{k=1}^{\infty} (A_1^{1/2} z, (I - 2B)^{-1} f_k)(A_1^{1/2} z, f_k)\right\} =$$
$$= \exp\left\{-\frac{1}{2} (A_1^{1/2} z, (I - 2B)^{-1} A_1^{1/2} z)\right\} =$$
$$= \exp\left\{-\frac{1}{2} (A_1^{1/2} (I - 2B)^{-1} A_1^{1/2} z, z)\right\}.$$

Hence,

$$A_2 = A_1^{1/2} (I - 2B)^{-1} A_1^{1/2}. \tag{21}$$

Note that

$$U = A_1^{-1/2} A_2 A_1^{-1/2} - I = (I - 2B)^{-1} - I = 2B\,(I - 2B)^{-1}$$

will also be a Hilbert-Schmidt operator, i.e., $\operatorname{tr} U U^* = \operatorname{tr} U^2 < \infty$. We have proved

Theorem 4. *In order for $\mu(0, A_1)$ to be equivalent to $\mu(0, A_2)$, it is necessary and sufficient that there exist a symmetric Hilbert-Schmidt operator U for which the operator $I + U$ is invertible and*

$$A_2 = A_1 + A_1^{1/2}\, U\, A_1^{1/2}; \tag{22}$$

then

$$\frac{d\mu(0, A_2)}{d\mu(0, A_1)} = \exp\left\{-\frac{1}{2}\sum_{k,j} (U\,[I + U]^{-1}\, e_k, e_j)\left[\frac{(x, e_k)\,(x, e_j)}{\sqrt{\lambda_k \lambda_j}} - \delta_{kj}\right] + \eta\right\},$$

where $\eta = \frac{1}{2}\sum_{k=1}^{\infty}\left[\log(1 + \gamma_k) - \frac{\gamma_k}{1 + \gamma_k}\right]$ and γ_k is the sequence of eigenvalues of the operator U.

§ 18. Absolute Continuity of Mixed Measures

Let $(X, \mathfrak{B})$ be a measurable Hilbert space and $(Y, \mathfrak{C})$ some measurable space. Assume that a measure ν is given on $(Y, \mathfrak{C})$, and that for each $y \in Y$ a measure μ^y is defined on $(X, \mathfrak{B})$ with the function $\mu^y(B)$ $\mathfrak{C}$-measurable in y for all $B \in \mathfrak{B}$. Set

$$\mu(B) = \int \mu^y(B)\, \nu(dy)\,. \tag{1}$$

It is easy to see that μ is a measure on $(X, \mathfrak{B})$. It will be normalized if μ^y and ν are normalized. Measures representable by Eq. (1) will also be called *measures obtained by mixing the measures* μ^y, or *mixed measures*. Consider two mixed measures

$$\mu^k(B) = \int \mu^{k,y}(B)\, \nu^k(dy)\,, \qquad k = 1, 2\,, \tag{2}$$

$\mu^{1,y}, \mu^{2,y}$ and ν^1, ν^2 satisfying the same conditions as μ^y and ν. The main goal of this section is the investigation of conditions for the absolute continuity of measures of the form (2).

We first consider the measures π^k defined on the product of measurable spaces $(X, \mathfrak{B}) \times (Y, \mathfrak{C})$ by the relation

$$\pi^k(B \times C) = \int_C \mu^{k,y}(B)\, \nu^k(dy)\,, \qquad k = 1, 2\,. \tag{3}$$

Since the measures μ^k are obtainable from the measures π^k by a transformation of the space $X \times Y$ into X by means of the function $f(x, y) = x$, in the investigation of the absolute continuity of measures of the form (2), we can use the results on the absolute continuity of those of the form (3).

It follows from the condition imposed on $\mu^{k,y}$ that for any $\mathfrak{B} \times \mathfrak{C}$-measurable bounded function $\varphi(x, y)$, the function $\int \varphi(x, y)\, \mu^{k,y}(dx)$ will be $\mathfrak{C}$-measurable and

$$\int \varphi(x, y)\, \pi^k(dx, dy) = \int \left[\int \varphi(x, y)\, \mu^{k,y}(dx)\right] \nu^k(dy) . \tag{4}$$

Theorem 1. *If the measures π^k are defined by (3), where $\mu^{k,y}$ and ν^k, $k = 1, 2$, satisfy the conditions enumerated above, then:* 1) *if $\pi^2 \ll \pi^1$, then $\nu^2 \ll \nu^1$ and for almost all y (mod ν^2), $\mu^{2,y} \ll \mu^{1,y}$;* 2) *conversely, if $\nu^2 \ll \nu^1$ and for almost all y (mod ν^2), $\mu^{2,y} \ll \mu^{1,y}$, then $\pi^2 \ll \pi^1$ and there exists a $\mathfrak{B} \times \mathfrak{C}$-measurable function $\tilde{\varrho}(x, y)$ such that for ν^2-almost all y*

$$\tilde{\varrho}(x,y) = \frac{d\mu^{2,y}}{d\mu^{1,y}}(x)$$

and

$$\frac{d\pi^2}{d\pi^1}(x, y) = \frac{d\nu^2}{d\nu^1}(y)\, \tilde{\varrho}(y, x) . \tag{5}$$

Proof. Assume $\pi^2 \ll \pi^1$ and set

$$\varrho(y, x) = \frac{d\pi^2}{d\pi^1}(x, y) .$$

Then for all $B \in \mathfrak{B}$ and $C \in \mathfrak{C}$ we have from (4):

$$\pi^2 (B \times C) = \int_C \int_B \varrho(y, x)\, \mu^{1,y}(dx)\, \nu^1(dy) =$$

$$= \int_C \int_B \frac{\varrho(y, x)\, \mu^{1,y}(dx)}{\int_X \varrho(y, x')\, \mu^{1,y}(dx')} \left[\int_X \varrho(y, x')\, \mu^{1,y}(dx')\right] \nu^1(dy) .$$

Putting $B = X$, we get

$$\pi^2(X \times C) = \nu^2 (C) = \int_C \left[\int_X \varrho(y, x)\, \mu^{1,y}(dx)\right] \nu^1(dy) ,$$

so that $\nu^2 \ll \nu^1$ and

$$\frac{d\nu^2}{d\nu^1}(y) = \int_X \varrho(y, x)\, \mu^{1,y}(dx) .$$

Consequently,

$$\pi^2 (B \times C) = \int_C \mu^{2,y}(B)\, \nu^2(dy) = \int_C \left[\int_B \tilde{\varrho}\, (y, x)\, \mu^{1,y}(dx)\right] \nu^2(dy) ,$$

where

$$\tilde{\varrho}(y, x) = \varrho(y, x) \left[\int_X \varrho(y, x)\, \mu^{1,y}(dx)\right]^{-1} . \tag{6}$$

Hence, for almost all y (mod ν^2) we have for each $B \in \mathfrak{B}$ the relation

$$\mu^{2,y}(B) = \int_B \tilde{\varrho}(y, x)\, \mu^{1,y}(dx) . \tag{7}$$

Choose a sequence of sets $B_k \in \mathfrak{B}$ in such a way that it forms the algebra generating $\mathfrak{B}$; this is possible because of the separability of $\mathfrak{B}$. We can then find a set $\tilde{C} \in \mathfrak{C}$ for which $\nu^2(\tilde{C}) = 1$ and for all $y \in \tilde{C}$ and all k the relation

$$\mu^{2,y}(B_k) = \int_{B_k} \tilde{\varrho}(y, x)\, \nu^{1,y}(dx)$$

holds. Then (7) is satisfied for all $y \in \tilde{C}$ and $B \in \mathfrak{B}$. Assertion (1) is proved. To prove (2) we first show that $\pi^2 \ll \pi^1$. Let $A \in \mathfrak{B} \times \mathfrak{C}$, $A_y = \{y\colon (x, y) \in A\}$. Then from (4)

$$\pi^k(A) = \int \mu^{k,y}(A_y)\, \nu^k(dy) .$$

If $\pi^1(A) = 0$, then $\mu^{1,y}(A_y) = 0 \pmod{\nu^1}$. Hence, $\mu^{2,y}(A_y) = 0 \pmod{\nu^2}$ since $\nu^2 \ll \nu^1$ and $\mu^{2,y} \ll \mu^{1,y}$ for almost all $y \pmod{\nu^2}$. Thus, $\pi^2(A) = 0$ so that $\pi^2 \ll \pi^1$. For the proof of the existence of $\tilde{\varrho}$ and the derivation of (5) we make use of the proof of 1), determining $\tilde{\varrho}$ from (6). □

Corollary. *If $\nu^2 \ll \nu^1$ and for ν^2-almost all y $\mu^{2,y} \ll \mu^{1,y}$, then $\mu^2 \ll \mu^1$.*

The density of the measure μ^2 w.r.t. μ^1 can be calculated with Theorem 6 (see (19)) § 15. Since in our case $f(x; y) = x$, we have

$$\frac{d\mu^2}{d\mu^1}(x) = \int \frac{d\nu^2}{d\nu^1}(y)\, \frac{d\mu^{2,y}}{d\mu^{1,y}}(x)\, \pi^1(dy/x) , \tag{8}$$

where $\pi^1(C/x)$ is a conditional measure defined by

$$\int_D \pi^1(C/x)\, \mu^1(dx) = \pi^1(B \times C) . \tag{9}$$

If $\mathfrak{A}^X$ and $\mathfrak{A}^Y$ are sub-algebras of $\mathfrak{B} \times \mathfrak{C}$, the former containing sets of the form $B \times Y$, $B \in \mathfrak{B}$ and the latter those of the form $X \times C$, $C \in \mathfrak{C}$, then $\pi(C/x)$ coincides with the value of the conditional measure π^1 w.r.t. the σ-algebra $\mathfrak{A}^X$ on the set $X \times C \in \mathfrak{A}^Y$.

The determination of the measure $\pi^1(C/x)$ from (9) is not always easy. We mention two cases in which $\frac{d\mu^2}{d\mu^1}$ can easily be expressed in terms of $\frac{d\nu^2}{d\nu^1}$ and $\frac{d\mu^{2,y}}{d\mu^{1,y}}$.

Lemma 1. *Assume the conditions of statement* 1) *of Theorem* 1 *are fulfilled. If, in addition:* a) *there exists a measure $\bar{\mu}$ for which $\mu^{1,y} \ll \bar{\mu}$ for all y, then*

$$\frac{d\mu^2}{d\mu^1}(y) = \int_Y \tilde{\varrho}(y, x)\, \frac{d\mu^{1,y}}{d\bar{\mu}}(x)\, \nu^2(dy) \Bigg/ \int_Y \frac{d\mu^{1,y}}{d\bar{\mu}}(x)\, \nu^1(dy) ; \tag{10}$$

if b) *all measures $\mu^{1,y}$ for differing y are mutually singular and there exist pairwise disjoint $B_y \in \mathfrak{B}$ for which $\mu^{1,y}(B_y) = 1$ and $\mu^{1,y}(B_{y'}) = 0$ when*

$y \neq y'$, the function

$$\varrho(x) = \frac{d\mu^{2,y}}{d\mu^{1,y}}(x), \qquad x \in B_y$$

is $\mathfrak{B}$-measurable and $\nu^1 = \nu^2$, then $\frac{d\mu^2}{d\mu^1}(x) = \varrho(x)$.

Proof. a) Let

$$\eta(x, y) = \frac{d\mu^{1,y}}{d\bar{\mu}}(x).$$

It is possible to choose $\eta(x, y)$ $\mathfrak{B} \times \mathfrak{C}$-measurable; it is sufficient to use Theorem 1 considering together with π^1 the measure $\bar{\pi} = \bar{\mu} \times \nu^1$ ($\pi^1 \ll \pi^2$). Then

$$\pi^1(B \times C) = \int_B \int_C \eta(x, y)\, \bar{\mu}(dx)\, \nu^1(dy),$$

and

$$\mu^1(B) = \pi^1(B \times Y) = \int_B \int_Y \eta(x, y)\, \nu^1(dy)\, \bar{\mu}(dx).$$

This implies that

$$\mu^1 \ll \bar{\mu}$$

and

$$\frac{d\mu^1}{d\bar{\mu}}(x) = \int_Y \eta(x, y)\, \nu^1(dy). \tag{11}$$

Obviously, $\mu^{2,y} \ll \bar{\mu}$ and

$$\frac{d\mu^{2,y}}{d\mu^{1,y}}(x) = \tilde{\varrho}(y, x)\, \eta(x, y).$$

In the same way as above, we obtain

$$\frac{d\mu^2}{d\bar{\mu}}(x) = \int_Y \tilde{\varrho}(y, x)\, \eta(x, y)\, \nu^2(dy). \tag{12}$$

Since for $\mu^2 \ll \mu^1 \ll \bar{\mu}$

$$\frac{d\mu^2}{d\mu^1} = \frac{d\mu^2}{d\bar{\mu}} \left(\frac{d\mu^1}{d\bar{\mu}}\right)^{-1},$$

we get (10) from (11) and (12).

To prove b) we have for any bounded $\mathfrak{B}$-measurable function $\varphi(x)$

$$\begin{aligned}\int \varphi(x)\, \varrho(x)\, \mu^1(dx) &= \int \left[\int \varphi(x)\, \varrho(x)\, \mu^{1,y}(dx)\right] \nu^1(dy) = \\ &= \int \left[\int \varphi(x) \frac{d\mu^{2,y}}{d\mu^{1,y}}(x)\, \mu^{1,y}(dx)\right] \nu^1(dy) = \\ &= \int \left[\int \varphi(x)\, \mu^{2,y}(dx)\right] \nu^1(dy) = \\ &= \int \varphi(x)\, \mu^2(dx).\end{aligned}$$

The lemma is proved. □

Assume Condition b) of Lemma 1 is satisfied for the measures $\mu^{1,y}$. Then the mapping $f((x;y)) = x$ admits μ^1-almost everywhere an inverse

$$(x;y) = g(x),$$

where $g(x)$ is defined for $x \in B_y$ as equal to $(x;y)$. Thus, $g(x)$ is defined on $\bigcup B_y$. This set need not be $\mathfrak{B}$-measurable but it will be $\overline{\mathfrak{B}}$-measurable, where $\overline{\mathfrak{B}}$ is the intersection of the completions of $\mathfrak{B}$ in the measures $\mu^{1,y}$ for all y. In particular, this set will be measurable w.r.t. the completion of μ^1 and its measure according to this completion equals 1. In order not to have to use completed measures and σ-algebras, we assume that the sets B_y are such that $\bigcup_{y\in C} B_y \in \mathfrak{B}$ for all $C \in \mathfrak{C}$. Then also $\bigcup_y B_y \in \mathfrak{B}$ and the function $g(x)$ is $\mathfrak{B}$-measurable:

$$\{x: g(x) \in B \times C\} = \bigcup_{y\in C} B_y \cap B \in \mathfrak{B}.$$

Consider the measure $\tilde{\pi}^1$ on $\mathfrak{B} \times \mathfrak{C}$ which results from μ^1 under the transformation g:

$$\begin{aligned}\tilde{\pi}^1(B \times C) &= \mu^1\Big(\bigcup_{y\in C} B_y \cap B\Big) = \int \mu^{1,y'}\Big(\bigcup_{y\in C} B_y \cap B\Big)\, \nu^1(dy') = \\ &= \int \mu^{1,y'}(B_{y'} \cap B)\, \chi_C(y')\, \nu^1(dy') = \\ &= \int \mu^{1,y'}(B)\, \chi_C(y')\, \nu^1(dy') = \pi^1(B \times C).\end{aligned}$$

We used the fact that

$$\mu^{1,y'}\Big(\bigcup_{y\in C} B_y \cap B\Big) = \mu^{1,y'}\Big(B_{y'} \cap \Big[\bigcup_{y\in C} B_y \cap B\Big]\Big) = \chi_C(y)\, \mu^{1,y'}(B),$$

where $\chi_C(y)$ is the indicator of the set C. Hence, if the measures $\mu^{1,y}$ satisfy the condition above with the same sets B_y, then from the fact that $\mu^2 \ll \mu^1$ it follows that $\pi^2 \ll \pi^1$. But then we can use Theorem 1. We have proved

Theorem 2. *Assume the family of measures $\mu^{k,y}$ on $(X, \mathfrak{B})$, $k = 1, 2$, satisfies:* 1) *$\mu^{k,y}(B)$ is $\mathfrak{C}$-measurable for all $B \in \mathfrak{B}$ and* 2) *there exist $\mathfrak{B}$-measurable sets B_y such that* $\mu^{k,y}(B_{y'}) = \begin{cases} 1; & y = y' \\ 0; & y \neq y' \end{cases}$, whereby $\bigcup_{y\in C} B_y \in \mathfrak{B}$ *for all $C \in \mathfrak{C}$, and $B_y \cap B_{y'} = \phi$ for $y \neq y'$. Also let ν^k, $k = 1, 2$, be measures on $(Y, \mathfrak{C})$. If the measures μ^k satisfy* (2), *then $\mu^2 \ll \mu^1$ iff* a) *$\nu^2 \ll \nu^1$ and* b) *ν^2-for almost all y $\mu^{2,y} \ll \mu^{1,y}$, whereby*

$$\frac{d\mu^2}{d\mu^1}(x) = \frac{d\mu^{2,y}}{d\mu^{1,y}}(x) \cdot \frac{d\nu^2}{d\nu^1}(x), \qquad x \in B_y. \tag{13}$$

To prove this we need only do (13). Introduce a new measure

$$\tilde{\mu}^1(B) = \int \mu^{1,y}(B)\, \nu^2(dy).$$

Then, on the basis of the assertion b) of Lemma 1

$$\frac{d\mu^2}{d\tilde{\mu}^1}(x) = \frac{d\mu^{2,y}}{d\mu^{1,y}} \qquad \text{for } x \in B_y .$$

Let us find $\frac{d\tilde{\mu}^1}{d\mu^1}$. For any bounded $\mathfrak{B}$-measurable function φ we have, setting $\frac{d\nu^2}{d\nu^1}(x) = \varrho(x)$ for $x \in B_y$,

$$\int \varphi(x)\, \tilde{\mu}^1(dx) = \int \Big[\int \varphi(x)\, \mu^{1,y}(dx) \Big] \nu^2(dy) =$$

$$= \int \Big[\int \varphi(x)\, \mu^{1,y}(dx) \Big] \frac{d\nu^2}{d\nu^1}(y)\, \nu^1(dy) =$$

$$= \int \Big[\int \varphi(x)\, \varrho(x)\, \mu^{1,y}(dx) \Big] \nu^1(dy) =$$

$$= \int \varphi(x)\, \varrho(x)\, \mu^1(dx) .$$

Thus, $\frac{d\tilde{\mu}^1}{d\mu^1}(x) = \varrho(x)$. To prove (13) it remains only to note that

$$\frac{d\mu^2}{d\mu^1} = \frac{d\mu^2}{d\tilde{\mu}^1} \cdot \frac{d\tilde{\mu}^1}{d\mu} .$$

We now give an example in which the result of Theorem 2 is used.

Example. Let $(Y, \mathfrak{C})$ be the positive real axis with σ-algebra of Borel sets, $\mu^{1,\eta}(\eta \in Y)$ a Gaussian measure with mean 0 and correlation operator A, $\mu^{2,\eta}$ a Gaussian measure with mean a and correlation operator A, ν some measure on $(Y, \mathfrak{C})$ and

$$\mu^k(B) = \int \mu^{k,\eta}(B)\, \nu(d\eta) .$$

To construct the sets B_η we introduce an orthonormalized sequence of eigenvectors $\{e_k\}$ of the operator and $\{\lambda_k\}$ the corresponding eigenvalues. We will show that the limit

$$\lim_{n\to\infty} \frac{1}{n} \sum_{k=1}^{n} \frac{(x, e_k)^2}{\lambda_k} = \eta \tag{14}$$

exists $\mu^{1,\eta}$ — a.e. From the properties of the Gaussian distribution (see (3) § 5) it is easy to deduce that the sequence of functions

$$\psi_n(x) = \sum_{k=1}^{n} \left[\frac{(x, e_k)^2}{\lambda_k} - \eta \right]$$

is a martingale w.r.t. the measure $\mu^{1,\eta}$ since for an arbitrary bounded Borel function $\varphi(t_1, \ldots, t_m)$ in R^m

$$\int \psi_{m+1}(x)\, \varphi((x, e_1), \ldots, (x, e_m))\, \mu^{1,\eta}(dx) =$$

$$= \int \psi_m(x)\, \varphi((x, e_1), \ldots, (x, e_m))\, \mu^{1,\eta}(dx) .$$

Hence, from Lemma 2

$$\mu^{1,\eta}\left(\left\{x\colon \sup_{k\le n} |\psi_n(x)| \ge a\right\}\right) \le \frac{1}{a}\int |\psi_n(x)|\, \mu^{1,\eta}(dx) \le$$

$$\leqq \frac{1}{a}\sqrt{\int \psi_n(x)^2\, \mu^{1,\eta}(dx)} =$$

$$= \frac{1}{a}\sqrt{\int \sum_{k,j=1}^{n} \left[\frac{(x, e_k)^2}{\lambda_k} - \eta\right]\cdot\left[\frac{(x, e_j)^2}{\lambda_j} - \eta\right] \mu^{1,\eta}(dx)} =$$

$$= \frac{\sqrt{n}}{a}\sqrt{\int \left[\frac{(x, e_1)^2}{\lambda_1} - \eta\right]^2 \mu^{1,\eta}(dx)}\,.$$

To prove (14) it is sufficient to show that for any $\varepsilon > 0$

$$\mu^{1,\eta}\left(\bigcap_{N=1}^{\infty}\bigcup_{n=N}^{\infty}\left\{x\colon \sup_{k\le 2^n} |\psi(x)| > \varepsilon\cdot 2^n\right\}\right) = 0\,.$$

But

$$\mu^{1,\eta}\left(\bigcup_{n=N}^{\infty}\left\{x\colon \sup_{k\le 2^n} |\psi_k(x)| > \varepsilon\cdot 2^n\right\}\right) \le \sum_{n=N}^{\infty} O\left(\frac{1}{\varepsilon}\cdot 2^{-n/2}\right) \to 0$$

for $N \to \infty$. (14) is proved. Let a be such that

$$\sum \frac{(a, e_k)^2}{\lambda_k} < \infty\,.$$

Then, for all η

$$\frac{d\mu^{2,\eta}}{d\mu^{1,\eta}}(x) = \exp\left\{\sum_{k=1}^{\infty} \frac{(a, e_k)\,(x, e_k)}{\eta\,\lambda_k} - \frac{1}{2}\sum_{k=1}^{\infty} \frac{(a, e_k)^2}{\eta\,\lambda_k}\right\}$$

so that

$$\frac{d\mu^2}{d\mu^1}(x) = \exp\left\{\sum_{k=1}^{\infty} \frac{(a, e_k)\,(x, e_k)}{\psi(x)\,\lambda_k} - \frac{1}{2}\sum_{k=1}^{\infty} \frac{(a, e_k)^2}{\psi(x)\,\lambda_k}\right\},$$

where

$$\psi(x) = \lim_{n\to\infty} \frac{1}{n}\sum_{k=1}^{n} \frac{(x, e_k)^2}{\lambda_k}\,.$$

Chapter 4. Admissible Shifts and Quasi-invariant Measures

§ 19. Admissible Shifts of Measures

In finite-dimensional spaces a very important role is played by Lebesgue measure, which remains invariant under all isometric transformations of such a space. In particular, it does not change under arbitrary translations of the space. As we will see in this section, there are no such measures in an infinite-dimensional Hilbert space. (We have in mind σ-finite measures). Moreover, there exists no σ-finite measure with the property that all measures obtained from it under all possible shifts of the space will be absolutely continuous w.r.t. to this original measure. In the study of the properties of measures an important place is occupied by transformations of a space which transfer a given measure into another measure absolutely continuous w.r.t. the original one. It is certainly quite reasonable to begin our study with shift transformations because of their simplicity. It will turn out that the study of shifts of a space under which a measure is sent into one absolutely continuous w.r.t. the original (such shifts are said to be *admissible*) makes it possible to investigate more complicated measure transformations. This chapter is devoted to the study of admissible shifts and in this section we will consider their general properties.

Let $S_a x = x + a$, i.e., S_a is translation (by a) in X. For all $A \in \mathfrak{B}$ we set $S_a A = \{x : x - a \in A\}$. If a measure μ is given on $(X, \mathfrak{B})$, then μ_a denotes the measure on $(X, \mathfrak{B})$ defined by the relation

$$\mu_a(A) = \mu(S_{-a} A) .$$

Definition. An element $a \in X$ is called an *admissible shift* for the measure μ if

$$\mu_a \ll \mu .$$

The set of admissible shifts of the measure μ will be denoted by M_μ. For all $a \in M_\mu$ we define the function $\varrho(a, x)$ as

$$\varrho(a, x) = \frac{d\mu_a}{d\mu}(x) . \quad (1)$$

We note some properties of the set M_μ and the function $\varrho(a, x)$.

I. The set M_μ is an additive semi-group and for $a \in M_\mu$, $b \in M_\mu$

$$\varrho(a + b, x) = \varrho(a, x) \cdot \varrho(b, x - a) . \tag{2}$$

Indeed, for any bounded $\mathfrak{B}$-measurable function $f(x)$ $\int f(x)\, \mu_{a+b}(dx) = = \int f(x + a + b)\, \mu(dx) = \int f(x + a)\, \varrho(b, x)\, \mu(dx) = \int f(x)\, \varrho(b, x - a) \times \times \varrho(a, x)\, \mu(dx)$. It follows from this relation that $\mu_{a+b} \ll \mu$ and (2) holds.

II. If $\varrho(a, x) > 0$ (mod μ), then $\mu_a \sim \mu$ and $-a \in M_\mu$, and

$$\varrho(-a, x) = \frac{1}{\varrho(a, x - a)} . \tag{3}$$

The fact that $\mu_a \sim \mu$ for $\varrho(a, x) > 0$ (mod μ) follows from the general properties of densities of measures formulated in § 13. Since the measures μ and μ_{-a} are obtained from equivalent measures under a one-to-one transformation of $(X, \mathfrak{B})$, they are also equivalent. Hence $-a \in M_\mu$. Formula (3) follows from (2) if we set $b = -a$ and take into account the fact that $\varrho(0, x) = 1$.

III. If $\nu \ll \mu$ and $\frac{d\nu}{d\mu}(x) = g(x)$, then

$$M_\mu \cap M_\nu = M_\mu \cap \{a : \mu(\{x : g(x) = 0, g(x - a)\, \varrho_\mu(a, x) > 0\}) = 0\}$$

and

$$\varrho_\nu(a, x) = \frac{g(x - a)}{g(x)}\, \varrho_\mu(a, x) \qquad (\text{mod } \nu) . \tag{4}$$

Here,

$$\varrho_\mu = \frac{d\mu_a}{d\mu} \quad \text{and} \quad \varrho_\nu = \frac{d\nu_a}{d\nu} ,$$

If f is a nonnegative $\mathfrak{B}$-measurable function, then for $a \in M_\mu$

$$\int f(x + a)\, \nu(dx) = \int f(x + a)\, g(x)\, \mu(dx) = \\ = \int f(x)\, g(x - a)\, \varrho_\mu(a, x)\, \mu(dx) . \tag{5}$$

If $g(x - a)\, \varrho_\mu(a, x) = 0$ for μ-almost all x for which $g(x) = 0$, then

$$\int f(x + a)\, \nu(dx) = \int f(x)\, \frac{g(x - a)\, \varrho_\mu(a, x)}{g(x)}\, g(x)\, \mu(dx) = \\ = \int f(x)\, \frac{g(x - a)\, \varrho_\mu(a, x)}{g(x)}\, \nu(dx)$$

so that $a \in M_\nu$ and (4) is established. Now let $a \in M_\mu \cap M_\nu$. Then from (5)

$$\int f(x)\, \varrho_\nu(a, x)\, g(x)\, \mu(dx) = \int f(x)\, g(x - a)\, \varrho_\mu(a, x)\, \mu(dx) .$$

Hence,

$$\varrho_\nu(a, x)\, g(x) = g\,(x - a)\, \varrho_\mu(a, x) \qquad (\text{mod}\, \mu)\,.$$

and $g(x - a)\, \varrho_\mu(a, x) = 0$ for μ-almost all x for which $g(x) = 0$, i.e.,

$$\mu\,(\{x\colon g(x) = 0, g(x - a)\, \varrho_\mu(a, x) > 0\}) = 0\,.$$

IV. If $\nu \sim \mu$, then $M_\mu = M_\nu$ and (4) holds. Indeed, in this case $\mu\,(\{x\colon g(x) = 0\}) = 0$ so that $M_\mu \subset M_\nu$ follows from III. Exchanging ν and μ, we obtain $M_\nu \subset M_\mu$.

We will call a vector $a \in X$ a *partially admissible shift* for the measure μ if $\frac{d\mu_a}{d\mu}$ is not identically zero (mod μ), i.e., if μ_a contains a component absolutely continuous w.r.t. μ. For partially admissible shifts we will write

$$\tilde\varrho_\mu(a, x) = \frac{d\mu_a}{d\mu}\,(x)$$

(we take here the density of the absolutely continuous component), and denote by $\widetilde M_\mu$ the set of partially admissible shifts. Clearly, $M_\mu \subset \widetilde M_\mu$ and for $a \in M_\mu$, the functions ϱ and $\tilde\varrho$ coincide.

A vector a is called an *inadmissible shift* if $\mu_a \perp \mu$.

V. Let $\nu \ll \mu$ and $\frac{d\nu}{d\mu}\,(x) = g(x)$. We have $a \in M_\nu$ only if $a \in \widetilde M_\mu$, whereby

$$\varrho_\nu(a, x) = \frac{g(x - a)}{g(x)}\, \tilde\varrho_\mu(a, x) \qquad (\text{mod}\, \nu)\,. \tag{6}$$

We note that if a is an inadmissible shift for the measure μ, then it will also be inadmissible for the measure ν. Hence $M_\nu \subset \widetilde M_\mu$. Further

$$\int f(x)\, \nu_a(dx) = \int f(x)\, \varrho_\nu(a, x)\, \nu(dx) = \int f(x)\, \varrho_\nu(a, x)\, g(x)\, \mu(dx)\,;$$

$$\begin{aligned}\int f(x + a)\, \nu(dx) &= \int f(x + a)\, g(x)\, \mu(dx) = \int f(x)\, g(x - a)\, \mu_a(dx) = \\ &= \int f(x)\, g(x - a)\, \tilde\varrho_\mu(a, x)\, \mu(dx) + \int f(x)\, g(x - a)\, \mu_a'(dx)\,,\end{aligned}$$

where $\mu_a' \perp \mu$, so that also $\nu \perp \mu_a'$ and $\nu_a \perp \mu_a'$. If $f = 0$ (mod μ_a'), then from

$$\int f(x)\, \varrho_\nu(a, x)\, g(x)\, \mu(dx) = \int f(x)\, g(x - a)\, \tilde\varrho_\mu(a, x)\, \mu(dx)$$

there also follows (6).

VI. Let $(X, \mathfrak{B})$ and $(Y, \mathfrak{C})$ be two Hilbert spaces and assume a measure μ is defined on $(X, \mathfrak{B})$. Let $y = T\,x$ be a linear map of X into Y and denote by ν the measure defined by $\nu(C) = \mu(T^{-1}\,C)$. If $a \in M_\mu$, then $b = T\,a \in M_\mu$ and

$$\varrho_\nu(b, y) = \int \varrho_\mu(a, x)\, \mu(dx, \mathfrak{B}/T^{-1}\,y)\,,$$

where $\mu(\cdot, \mathfrak{B}/y)$ is a conditional distribution for the measure μ w.r.t. the σ-algebra $\mathfrak{B}$ generated by sets of the form $T^{-1}C$, $C \in \mathfrak{C}$. In particular, if the map T is invertible, then

$$\varrho_\nu(b, y) = \varrho_\mu(T^{-1}b, T^{-1}y) .$$

This assertion follows immediately from Theorem 6 § 15.

VII. Let the measures μ_1 and μ_2 be given on $(X, \mathfrak{B})$ and assume $\mu = \mu_1 * \mu_2$ is their convolution. Then for $a \in M_{\mu_1}$ we have $a \in M_\mu$ and

$$\varrho_\mu(a, x) = \int \varrho_{\mu_1}(a, y)\, \mu(dy \times X, \mathfrak{B}_0/U^{-1}x) , \tag{7}$$

where $\mu(dy, \mathfrak{B}_0/x)$ is the conditional measure for $\mu_1 \times \mu_2$ w.r.t. the σ-algebra $\mathfrak{B}_0 \subset \mathfrak{B} \times \mathfrak{B}$ generated by sets of the form $\{(x_1; x_2); x_1 + x_2 \in \in A\}$, $A \subset \mathfrak{B}$ and U is a mapping of $X \times X$ into X: $U((x_1; x_2)) = = x_1 + x_2$. Formula (7) also follows easily from Theorem 6 and its consequences.

VIII. We consider a measure μ on $(X, \mathfrak{B})$ and introduce in M_μ the following distance between two elements:

$$r(a_1, a_2) = \int |\varrho_\mu(a_1, x) - \varrho_\mu(a_2, x)|\, \mu(dx) . \tag{8}$$

Then M_μ with the distance r is a complete metric space. Let $a_n \in M_\mu$ and

$$\lim_{n\to\infty,\, m\to\infty} r(a_n, a_m) = 0 .$$

We will show that there exists an $a \in M_\mu$ for which $\lim\limits_{n\to\infty} r(a_n, a) = 0$. The sequence a_n is bounded. Indeed, when $|a_{n_k}| \to \infty$ and we choose n_k so that $r(a_{n_k}, a_{n_{k+1}}) < \frac{1}{3^k}$, then for a $\mathfrak{B}$-measurable finite function f with $\|f\| = 1$ we will have

$$\int f(x + a_{n_k})\, \mu(dx) \geq \int f(x + a_{n_1})\, \mu(dx) - \sum_{k=1}^{N-1} r(a_{n_k}, a_{n_{k+1}}) .$$

Choosing f so that $\int f(x + a_{n_1})\, \mu(dx) > \frac{1}{2}$ and letting $N \to \infty$, we obtain

$$\varlimsup_{N\to\infty} \int f(x + a_{n_N})\, \mu(dx) > 0 ,$$

which is not possible for finite f if $|a_{n_N}| \to \infty$. Hence, a_n is bounded. We can therefore assume without loss of generality that a_n converges weakly to some a. We will show that $a \in M_\mu$ and will find $\varrho_\mu(a, x)$. From the relation

$$\int e^{i(z, x+a_n)}\, \mu(dx) = e^{i(z, a_n)} \int e^{i(z,x)}\, \mu(dx) = \int e^{i(z,x)}\, \varrho_\mu(a_n, x)\, \mu(dx) ,$$

and also taking into account the fact that $(z, a_n) \to (z, a)$ and the existence (mod μ) of

$$\lim_{n\to\infty} \varrho_\mu(a_n, x) = \varrho(x)\,,$$

for which

$$\lim_{n\to\infty} \int |\varrho_\mu(a_n, x) - \varrho(x)|\mu(dx) = 0\,,$$

we find that

$$e^{i(z,a)} \int e^{i(z,x)}\, \mu(dx) = \int e^{i(z,x)}\, \varrho(x)\, \mu(dx)\,.$$

Thus,

$$\int e^{i(z,x)}\, \mu_a(dx) = \int e^{i(z,x)}\, \varrho(x)\, \mu(dx)$$

for all $z \in X$, so that $\mu_a \ll \mu$ and $\varrho(x) = \frac{d\mu_a}{d\mu}(x)$. The assertion is proved.

From the remark in § 4 it follows that for each measure μ we can find a nuclear operator B such that for the characteristic functional $\theta(z)$ of the measure μ

$$\operatorname{Re}\big(1 - \theta(z)\big) \to 0 \qquad \text{when } (B\,z, z) \to 0\,. \tag{9}$$

Theorem 1. *If $\theta(z)$ is the characteristic functional of the measure μ and the nuclear operator B is such that* (9) *is satisfied, then $\widetilde{M}_\mu \subset B^{1/2} X$ ($B^{1/2}$ is a nonnegative symmetric operator, the square root of B).*

Proof. Let $a \in \widetilde{M}_\mu$ and denote by μ^1 the component of μ for which μ_a^1 is absolutely continuous w.r.t. μ. Then

$$\operatorname{Re} \int [1 - \cos(z, x)]\, \mu^1(dx) \le \int [1 - \cos(z, x)]\, \mu(dx) = \operatorname{Re}\big(1 - \theta(z)\big)$$

which means that

$$\operatorname{Re} \int [1 - \cos(z, x)]\, \mu^1(dx) \to 0$$

when $(B\,z, z) \to 0$. Further

$$e^{i(z,a)} \int e^{i(z,x)}\, \mu^1(dx) = \int e^{i(x+a,z)}\, \mu^1(dx)$$

so that

$$\operatorname{Re}\big(\mu^1(X) - e^{i(z,a)} \int e^{i(z,x)}\, \mu^1(dx)\big) = \operatorname{Re} \int [1 - \cos(z, x + a)]\, \mu^1(dx) \le$$
$$\le \int [1 - \cos(z, x)\,]\, \varrho_\mu(a, x)\, \mu(dx)\,.$$

We now show that

$$\int [1 - \cos(z, x)\,]\, \varrho_\mu(a, x)\, \mu(dx) \to 0$$

for $(B\,z, z) \to 0$. Taking $\alpha > 0$ arbitrary, we have

$$\int [1 - \cos(z, x)]\, \varrho_\mu(a, x)\, \mu(dx) \le \alpha \int [1 - \cos(z, x)]\, \mu(dx) +$$
$$+ 2 \int_{\{x:\varrho_\mu(a,x) > \alpha\}} \mu(dx)\,.$$

The first term tends to zero for $(B z, z) \to 0$ and the second can be made arbitrarily small by choosing α sufficiently large. Hence,

$$\mathrm{Re}\left(\mu^1(X) - e^{i(z,a)} \int e^{i(z,x)} \mu^1(dx)\right) \to 0 \qquad \text{for } (B z, z) \to 0 .$$

Moreover,

$$|\mu^1(X) - \int e^{i(z,x)} \mu^1(dx)|^2 \leq \int |1 - e^{i(z,x)}|^2 \mu^1(dx) \leq$$
$$\leq 2 \int \left(1 - \cos(z, x)\right) \mu^1(dx) \to 0$$

when $(B z, z) \to 0$. Hence, $e^{i(z,a)} \to 1$ for $(B z, z) \to 0$ or $(z, a) \to 0$ when $(B z, z) \to 0$. If $|(z, a)| < \varepsilon$ for $(B z, z) < \delta$, then

$$(z, a)^2 < \frac{\varepsilon^2}{\delta} (B z, z) .$$

Hence, it follows that $(a, z) \to 0$ if $(B z, z) \to 0$, i.e., a belongs to the closure of $B^{1/2} X$. Then the linear functional $(a, B^{-1/2} z)$ is defined for $z \in B^{1/2} X$ as follows: if $z = B^{1/2} y$, then $(a, B^{-1/2} z) = (a, y)$ (for a from the closure of $B^{1/2} X$ this definition is unique since it follows from $z = B^{1/2} y_1 = B^{1/2} y_2$ that $B^{1/2}(y_2 - y_1) = 0$ so that $\left(B(y_1 - y_2), (y_1 - y_2)\right) = 0$ and $(a, y_1) = (a, y_2)$). This functional is bounded on $B^{1/2} X$:

$$|(a, B^{-1/2} z)|^2 \leq |(a, y)|^2 < \frac{\varepsilon^2}{\delta} (B y, y) = \frac{\varepsilon^2}{\delta} (B^{1/2} y, B^{1/2} y) = \frac{\varepsilon^2}{\delta} |z|^2 .$$

Therefore, it can be extended by continuity to the closure of $B^{1/2} X$ and is representable as

$$(a, B^{-1/2} z) = (b, z) .$$

We thus easily find that $a = B^{1/2} b$. □

Corollary. *For any infinite-dimensional subspace $L \subset X$, the set $L \cap \widetilde{M}_\mu$ is of the first category in X.* Indeed, one can find a nuclear operator B for which $L \cap \widetilde{M}_\mu \subset L \cap B(X)$. Let S be a closed sphere in X. We show that the set $L \cap B(S)$ is nowhere dense in L. This follows from the fact that $L \cap B(S)$ is a compact set: since a sphere in L is not compact, in each sphere of L one can find a point which does not belong to $L \cap B(S)$ and because the latter set is closed, this point can be enclosed by some sphere having no points in common with $L \cap B(S)$. If S_n is a closed sphere in X with center at 0 and radius n, then $L \cap B(S_n)$ is nowhere dense in L and

$$L \cap B(X) = \bigcup_n L \cap B(S_n) .$$

Our assertion follows from this formula. □

We now show that there exists no σ-finite measure in X for which each shift from X is admissible (in particular, there exists no σ-finite measure invariant w.r.t. all shifts of the space). If ν were such a measure, then we could define an everywhere positive function $\varphi(x)$ such that the measure

$$\mu(A) = \int_A \varphi(x)\, \nu(dx)$$

would be finite. Clearly, any admissible shift for the measure ν will also be admissible for the measure μ. But $\widetilde{M}_\mu$ cannot coincide with X (since X is of the second category and $\widetilde{M}_\mu$ is of the first).

Let us consider some other properties of the set $\widetilde{M}_\mu$. We will prove that $\widetilde{M}_\mu$ is a $\mathfrak{B}$-measurable set. Indeed, let $\widetilde{\mathfrak{B}}_n$ be a sequence of σ-algebras satisfying the conditions: a) $\widetilde{\mathfrak{B}}_n \subset \widetilde{\mathfrak{B}}_{n+1}$, b) $\bigcup \widetilde{\mathfrak{B}}_n$ generates the σ-algebra $\mathfrak{B}$ and c) each of the σ-algebras $\widetilde{\mathfrak{B}}_n$ is generated by a countable collection of pairwise disjoint $\mathfrak{B}$-measurable sets $(A_1^n, A_2^n, \ldots)$. If μ and ν are measures and $\tilde{\mu}^n$ and $\tilde{\nu}^n$ are the restrictions of these measures to $\widetilde{\mathfrak{B}}_n$, then $\dfrac{d\tilde{\nu}^n}{d\tilde{\mu}^n}$ is a martingale on the measurable space $(X, \mathfrak{B}, \mu)$ and

$$\lim_{n\to\infty} \frac{d\tilde{\nu}^n}{d\tilde{\mu}^n}(x) = \frac{d\nu}{d\mu}(x) \qquad (\operatorname{mod} \mu)\,.$$

This is proved in exactly the same way as in § 15 (see Theorem 3). Setting $\tilde{\varrho}_\mu(a, x) = 0$ for $a \notin \widetilde{M}_\mu$, we have

$$\tilde{\varrho}_\mu(a, x) = \lim_{n\to\infty} g_n(x, a) \qquad (\operatorname{mod} \mu)\,,$$

where

$$g_n(x, a) = \sum_{k=1}^{\infty} \frac{\mu_a(A_k^n)}{\mu(A_k^n)}\, \chi_{A_k^n}(x)\,;$$

when $\mu(A_k^n) = 0$ we take $\mu_a(A_k^n)/\mu(A_k^n) = 0$. It is clear that for $0 < \alpha < 1$

$$\int [\tilde{\varrho}_\mu(a, x)]^\alpha\, \mu(dx) = \lim_{n\to\infty} \int [g_n(x, a)]^\alpha\, \mu(dx)\,;$$

the possibility of going to the limit under the integral sign follows from the inequality

$$\int ([g_n(x, a)]^\alpha)^{1/\alpha}\, \mu(dx) = \int g_n(x, a)\, \tilde{\mu}^n(dx) \le \mu(X)\,.$$

Since $\mu_a(A_k^n)$ is a $\mathfrak{B}$-measurable function of $a \in X$,

$$\int [g_n(x, a)]^\alpha\, \mu(dx) = \sum_{k=1}^{\infty} [\mu_a(A_k^n)]^\alpha\, [\mu(A_k^n)]^{1-\alpha}$$

is also $\mathfrak{B}$-measurable w.r.t. a. Hence, $\int [\tilde{\varrho}_\mu(a, x)]^\alpha \mu(dx)$ is $\mathfrak{B}$-measurable w.r.t. a as the limit of $\mathfrak{B}$-measurable functions. Since

$$\tilde{M}_\mu = \{a: \int [\tilde{\varrho}_\mu(a, x)]^\alpha \mu(dx) > \alpha\} ,$$

$\tilde{M}_\mu$ is a $\mathfrak{B}$-measurable set. It is easy to see that the function $\int \tilde{\varrho}_\mu(a, x) \mu(dx)$ is $\mathfrak{B}$-measurable w.r.t. a, so that

$$M_\mu = \{a: \int \tilde{\varrho}_\mu(a, x) \mu(dx) = 1\}$$

is also $\mathfrak{B}$-measurable. Thus, it makes sense to consider $\mu(M_\mu)$. In the finite-dimensional case this quantity can assume any value in the interval $[0, \mu(X)]$. A measure in a Hilbert space X can be concentrated on a finite-dimensional subspace and we have an illustration of this here. However, in the important infinite-dimensional case the situation changes.

Theorem 2. *If for any finite-dimensional subspace $L \subset X$ $\mu(L) = 0$, then*

$$\mu(M_\mu) = 0 .$$

Proof. Assume the contrary holds: $\mu(M_\mu) > 0$. Let $\tilde{\mu}(A) = \mu(-A)$, where $-A = \{x: -x \in A\}$. Then $M_{\tilde{\mu}} = -M_\mu$ and $M_{\mu * \tilde{\mu}} \supset M_\mu + M_{\tilde{\mu}}$, $\mu * \tilde{\mu}(M_{\mu * \tilde{\mu}}) > \mu(M_\mu)\, \tilde{\mu}(M_{\tilde{\mu}}) = [\mu(M_\mu)]^2 > 0$. Thus, without loss of generality we can assume that the measure μ is symmetric and that M_μ is an additive group. Find a symmetric compact $K' \subset M_\mu$ for which $\mu(K') > 0$. Let Y be the linear hull of K' and K the convex closure of K' (K is a closed set). If we introduce a norm into Y in such a way that K becomes the unit sphere, then Y will be a complete normed space. From the hypotheses of the theorem it follows that it is infinite-dimensional. We now show that it is separable. Set $M' = Y \cap M_\mu$. Since Y is the convex closure of M' it suffices to show that M' contains a countable ε-net for all $\varepsilon > 0$. If this were not so, there would exist an uncountable set $Q \subset M'$ such that for $y_1, y_2 \in Q$,

$$\|y_1 - y_2\|_K \geq \varepsilon .$$

For $y \in Q$ set

$$V_n(y) = \{x \in Y: \|x - n y\|_K \leq 1\} ,$$

where n is such that $n\varepsilon > 2$. Then $ny \in M'$ and

$$\mu(V_n(y)) = \mu_{ny}(K) > 0$$

since $\mu(K) > 0$ and for $y_1 \neq y_2$ we have $V_n(y_1) \cap V_n(y_2) = \phi$ because $n\|y_1 - y_2\|_K > 2$. In this way one would construct an uncountable family of pairwise disjoint sets of positive measure which is impossible. This proves the separability of Y.

Denote by μ' the restriction of the measure μ to the σ-algebra $\mathfrak{B}_Y$ of Borel sets of Y. It is easy to see that $M_{\mu'} \supset M'$. Choose a sequence of compacts C_n in Y in such a way that

$$\lim_{n\to\infty} \mu(C_n) = \mu(Y) .$$

If Y_1 is the linear hull of $\bigcup_n C_n$, then $\mu'(Y_1) = \mu'(Y)$. Note that $Y_1 \neq Y$ since Y is an infinite-dimensional Banach space and cannot be represented as a countable sum of compacts. Hence $M' - Y_1$ is not empty. Let $a \in M' - Y_1$, $Y_2 = \{x : x - a \in Y_1\}$. Then $Y_1 \cap Y_2 = \phi$ so that $\mu'(Y_2) = 0$. But $\mu'_a(Y_2) = \mu'(Y_1) = \mu^1(X)$, i.e., $\mu'_a \perp \mu'$. We have arrived at a contradiction. □

Corollary. *If for some measure μ, $\mu(M_\mu) = \mu(X)$, then there exists a sequence of finite-dimensional subspaces L_n such that $\mu(X) = \mu\Big(\bigcup_n L_n\Big)$.*

Indeed, the set of finite-dimensional subspaces L for which $\mu(L) > 0$ is at most countable. If we let L_n be all such spaces, then setting

$$\mu^1(A) = \mu\left(A \cap (\bigcup_n L_n)\right), \qquad \mu^2(A) = \mu(A) - \mu^1(A)$$

and using the fact that $\mu^1_a \perp \mu^2$ for all a, we find that $M_\mu \subset M_{\mu^1}$. This means that $\mu^2(M_{\mu^1}) = \mu^2(X)$. Hence, from Theorem 2 $\mu^2(X) = 0$ so that

$$\mu(A) = \mu\left(A \cap [\bigcup_n L_n]\right).$$

As an example of a measure for which $\mu(M_\mu) = \mu(X)$ we can take a measure concentrated on the set of points $C = \Big\{x : x = \sum_k n_k x_k\Big\}$, where the n_k are integers and $\{x_k\}$ is an orthonormalized sequence, provided that the measure of all one-point sets in C is positive.

§ 20. Admissible Directions

A vector $a \in X$ for which $|a| = 1$ will be called an *admissible direction* for the measure μ if $\lambda a \in M_\mu$ for all $\lambda > 0$. The fact that a is an admissible shift for the measure μ says little about the structure of the measure. However, useful information on the structure of μ can be extracted from the fact that a is an admissible direction.

Denote by L_1 a one-dimensional space containing the vector a and by X^1 the orthogonal complement of L_1 in X. Moreover, let $\mathfrak{B}^1$ be the σ-algebra of Borel sets in X^1 and $\tilde{\mathfrak{B}}^1$ the σ-algebra of subsets of X of the form $\{x : P_{X^1} x \in A^1\}$, where $A^1 \subset \mathfrak{B}^1$ and P_{X^1} is projection on X^1. $\mathfrak{B}_1$ will denote the σ-algebra of cylinder sets with bases in L_1

and $\mu(\cdot, \mathfrak{B}_1/x)$ the conditional measure for μ w.r.t. the σ-algebra $\mathfrak{B}_1$, i.e., the function defined by

$$\mu(B \cap B_1) = \int_{B_1} \mu(B, \mathfrak{B}_1/x)\, \mu(dx)$$

for $B \subset \mathfrak{B}$ and $B_1 \subset \mathfrak{B}_1$. Since the function $\mu(\cdot, \mathfrak{B}_1/x)$ is $\mathfrak{B}_1$-measurable and each $\mathfrak{B}_1$-measurable function is a function of $P_{L_1} x$, we have

$$\mu(\cdot, \mathfrak{B}_1/x) = \mu(\cdot, \mathfrak{B}_1/(x, a)a) \qquad (\mathrm{mod}\, \mu) .$$

Define a measure μ^1 on L_1 as the projection of μ on L_1. For μ^1-almost all $x \in L_1$, the measure $\mu(x, B^1) = \mu(P_{X^1}^{-1} B^1, \mathfrak{B}_1/x)$ is defined on $\mathfrak{B}_1$. In what follows we will consider X as the Cartesian product $L_1 \times X^1$. The collection of measures $\mu(X, B^1)$ is uniquely $(\mathrm{mod}\, \mu^1)$ defined by the relation

$$\mu(E_1 \times B^1) = \int_{E_1} \mu(X, B^1)\, \mu^1(dx) , \tag{1}$$

where $B^1 \in \mathfrak{B}^1$ and $E_1 \in \mathfrak{B}_{L_1}$, and $\mathfrak{B}_{L_1}$ is the σ-algebra of Borel sets of L_1. It is obvious that for $\lambda > 0$

$$\mu_{\lambda a}(E_1 \times B^1) = \mu(\{E_1 - \lambda a\} \times B^1) = \int_{\{E_1 - \lambda a\}} \mu(x, B^1)\, \mu^1(dx) =$$

$$= \int_{E_1} \mu(x - \lambda a, B^1)\, \mu^1_{\lambda a}(dx) , \tag{2}$$

where $\{E - \lambda a\} = \{x : x + \lambda a \in E_1\}$. From Theorem 1, § 18 it follows that for all $\lambda > 0$ $\mu^1_{\lambda a} \ll \mu^1$ and also, for $\mu^1_{\lambda a}$-almost all $x \in L_1$

$$\mu(x_1 - \lambda a, \cdot) \ll \mu(x_1, \cdot) . \tag{3}$$

The problem of the determination of those measures on the line for which all positive shifts are admissible is solved by the following simple lemma, which we give without proof.

Lemma 1. *Let a measure μ on R^1 be such that $M_\mu \supset [0, \infty)$. Then $\mu \ll m$, where m is Lebesgue measure and there exists an s (possibly $s = -\infty$) for which $\frac{d\mu}{dm}(a) = 0$ for $a < s$ and $\frac{d\mu}{dm}(a) > 0$ for $a > s$.*

Denote by $\tilde{\mu}^1$ the projection of μ on X and let $\tilde{\mu} = \mu^1 \times \tilde{\mu}^1$ be a measure on $\mathfrak{B}$ defined by

$$\tilde{\mu}(E_1 \times B^1) = \mu^1(E_1)\, \tilde{\mu}^1(B^1) , \qquad E_1 \in \mathfrak{B}_{L_1},\ B^1 \in \mathfrak{B}^1 . \tag{4}$$

Theorem 1. *$\lambda a \in M_\mu$ for all $\lambda > 0$ iff the measure μ is absolutely continuous w.r.t. the measure $\tilde{\mu}$ defined by (4); the measure μ^1 is absolutely continuous w.r.t. Lebesgue measure on L_1; there exists a λ_1 (possibly $\lambda_1 = -\infty$) for which the density of μ^1 w.r.t. Lebesgue measure on L_1 is positive for $(a, x) > \lambda_1$ and the density $g(x) = \frac{d\mu}{d\tilde{\mu}}(x)$ satisfies for $\lambda > 0$*

the condition

$$\mu(\{x: g(x - \lambda a) > 0, g(x) = 0\}) = 0 . \tag{5}$$

Proof. The sufficiency follows from the fact that for $\lambda > 0$

$$\tilde{\mu}_{\lambda a} = \mu^1_{\lambda a} \times \tilde{\mu}^1 \tag{6}$$

and $\mu^1_{\lambda a} \ll \mu^1$ so that $\lambda a \in M_{\tilde{\mu}}$ and from (5) and Property III, § 19, $\lambda a \in M_{\tilde{\mu}} \cap M_\mu$.

Necessity. We first show that $\mu \ll \tilde{\mu}$. For all $A \subset \mathfrak{B}$ denote by A^1_x, $x \in L_1$, the set of all $y \in X^1$ for which $(x; y) \in A$ (we again consider X as $L_1 \times X^1$), i.e., A^1_x is the section of A by the hyperplane parallel to X^1 and passing through x. Then $A^1_x \in \mathfrak{B}^1$,

$$\mu(A) = \int \mu(x, A^1_x)\, \mu^1(dx) , \tag{7}$$

and

$$\tilde{\mu}(A) = \int \tilde{\mu}^1(A^1_x)\, \mu^1(dx) . \tag{8}$$

Hence, to prove $\mu \ll \tilde{\mu}$ it is sufficient to show that for all x $\mu(x, \cdot) \ll \tilde{\mu}^1$. But

$$\tilde{\mu}^1(B^1) = \int \mu(x, B^1)\, \mu^1(dx) .$$

If $\tilde{\mu}^1(B^1) = 0$, then $\mu(x, B^1) = 0$ for μ^1-almost all x. Since the density of μ^1 w.r.t. Lebesgue measure on L_1 is positive at the point x for sufficiently large (a, x), there exists a λ large enough so that $\mu(\lambda a, B^1) = 0$. Using the fact that $\mu(\lambda' a, \cdot) \ll \mu(\lambda a, \cdot)$ for $\lambda' < \lambda$ on the basis of (3), we convince ourselves that $\mu(\lambda a, B^1) = 0$ for all λ. Hence, when $\tilde{\mu}(B^1) = 0$, $\mu(x, B^1) = 0$ for all $x \in L_1$, i.e., $\mu(x, \cdot) \ll \tilde{\mu}^1$. Then also $\mu \ll \tilde{\mu}$. We now calculate $\frac{d\mu}{d\tilde{\mu}}(x) = g(x)$. Using Theorem 1, § 18 again as well as (7) and (8) we get

$$g(x) = \frac{d\mu(P_{L_1} x, \cdot)}{d\tilde{\mu}} (P_{X^1} x) .$$

Since $\mu(P_{L_1} x - \lambda a, \cdot) \ll \mu(P_{L_1} x, \cdot)$ for $\lambda > 0$, we have

$$g(x-a) = \frac{d\mu(P_{L_1} x - \lambda a, \cdot)}{d\tilde{\mu}} (P_{L_1} x) = \frac{d\mu(P_{L_1} x - \lambda a, \cdot)}{d\mu(P_{L_1} x, \cdot)} (P_{L_1} x) g(x) \pmod{\mu} , \tag{9}$$

whence follows the validity of (5). □

Remark 1. Using Property III, § 19 we can determine $\varrho_\mu(a, x)$. Let $\sigma(x)$ be the density of μ^1 w.r.t. Lebesgue measure on L_1. Then

$$\frac{d\mu^1_{\lambda a}}{d\mu^1}(x) = \frac{\sigma(x - \lambda a)}{\sigma(x)} .$$

Hence, on the basis of (6)

$$\varrho_\mu(\lambda a, x) = \frac{\sigma(P_{L_1} x - \lambda a)}{\sigma(P_{L_1} x)} .$$

From (9) it follows that

$$\frac{g(x-a)}{g(x)} = \frac{d\mu(P_{L_1}x - \lambda\, a, \cdot)}{d\mu(P_{L_1}x)}\,(P_{L_1}x) \qquad (\mathrm{mod}\,\mu)\,.$$

Hence, using (4) § 19 one gets

$$\varrho_\mu(a, x) = \frac{d\mu(P_{L_1}x - \lambda a, \cdot)}{d\mu(P_{L_1}x, \cdot)}\,(P_{L_1}x)\,\frac{\sigma(P_{L_1}x - \lambda a)}{\sigma(P_{L_1}x)}\,. \tag{10}$$

Remark 2. It is easy to see that when $\lambda a \in M_\mu$ also for $\lambda < 0$, then $\mu \sim \tilde{\mu}$ and the number λ_1 mentioned in Theorem 1 is equal to $-\infty$. Using (10), (7) and (8) we easily see that

$$\frac{d\tilde{\mu}}{d\mu}(x) = \sigma(P_{L_1}x)\int\limits_{-\infty}^{\infty}\varrho_\mu(\tau a, x)\,d\tau\,. \tag{11}$$

What can be said about the measure μ if the set of admissible shifts contains some finite-dimensional subspace L? Choose in L an orthonormalized basis $\{e_k,\ k = 1, \ldots, m\}$ with m the dimension of L. On the basis of Remark 2, we deduce that $\mu \sim \mu^1 \times \tilde{\mu}^1$, where μ^1 is the projection of μ on the subspace L_1 generated by the vector e_1 and $\tilde{\mu}^1$ is the projection of μ on the subspace X^1 orthogonal to L_1. Since for $a = \sum_{k=2}^{m} \lambda_k e_k$

$$[\mu^1 \times \tilde{\mu}^1]_a = \mu^1 \times \mu^1_a\,,$$

one has $a \in M_{\tilde{\mu}^1}$. Hence, the measure $\tilde{\mu}^1$ on X_1 will be equivalent to $\mu^2 \times \tilde{\mu}^2$, where μ^2 is the projection of $\tilde{\mu}^1$ on L_2, the subspace generated by the vector e_2, and $\tilde{\mu}^2$ is the projection of $\tilde{\mu}^1$ on X^2, the subspace of X^1 orthogonal to L_2. Here we consider X^1 as $L_2 \times X^2$ and X as $L_1 \times L_2 \times X^2$. Note that μ^2 and $\tilde{\mu}^2$ are projections of the same measure μ on L_2 and X^2, resp. Arguing in a similar way, we see that if we denote by L_k the subspace generated by e_k $(k = 1, \ldots, m)$, by X^m the subspace orthogonal to L_m, by μ^k the projection of μ on L_k and by $\tilde{\mu}^m$ the projection of μ on X^m, then we will have

$$\mu \sim \mu^1 \times \cdots \times \mu^m \times \tilde{\mu}^m\,, \tag{12}$$

whereby the space X can be considered as the Cartesian product $L_1 \times \cdots \times L_m \times X^m$. Each of the measures μ^k is equivalent to Lebesgue measure on L_k. In particular, the projection of μ on L will be equivalent to $\mu^1 \times \cdots \times \mu^m$ if we assume that $L = L_1 \times \cdots \times L_m$. Thus, the projection of μ on L will also be equivalent to Lebesgue measure on L.

Assume that M_μ contains N, the infinite-dimensional linear manifold of admissible shifts. Then, from the above it follows that for any finite-dimensional subspace $L \in N$, the measure μ_L — the projection of μ on L — is equivalent to Lebesgue measure on L, but if $\tilde{\mu}_L$ is the projection of μ on the orthogonal complement of L, then

$$\mu \sim \mu_L \times \tilde{\mu}_L \,. \tag{13}$$

If N is everywhere dense in X, then we can choose an orthonormalized basis $\{e_k, k = 1, \ldots\}$ in N and for all m (11) will hold. It is natural to suppose that μ will be equivalent to the product measure

$$\mu^1 \times \cdots \times \mu^m \times \cdots = \prod_{k=1}^{\infty} \mu^k = \bar{\mu} \,.$$

However, this is not correct. To see this we note that for any $a \in X$, either $\bar{\mu}_a \sim \bar{\mu}$ or $\bar{\mu}_a \perp \bar{\mu}$ (this follows from Lemmas 1 and 2 § 16). Hence, for any measure μ equivalent to a product measure and also for all a, the following alternatives hold: either $\mu_a \sim \mu$ or $\mu_a \perp \mu$. We will give an example of a measure for which M_μ contains a linear manifold N everywhere dense in X and for which there exists at the same time an a such that μ_a is neither equivalent nor orthogonal to μ.

Let μ^1 and μ^2 be Gaussian measures with mean 0 and correlation operators B_1 and B_2. Choose B_1 and B_2 so that $B_1^{1/2} X \supset B_2^{1/2} X$ but $B_1^{1/2} X \neq B_2^{1/2} X$. If B_2 is a nonsingular operator, then $B_2^{1/2} X$ is everywhere dense in X. From the fact that $B_1^{1/2} X \neq B_2^{1/2} X$ it follows that $\mu^1 \perp \mu^2$ since otherwise we would have $\mu^1 \sim \mu^2$ (see Theorem 2, § 17) and $M_{\mu^2} = B_2^{1/2} X$, $M_{\mu^1} = B_1^{1/2} X$ (see Theorem 1, § 17). Let $\mu = \frac{1}{2}(\mu_1 + \mu_2)$. Then $M_\mu \supset B_2^{1/2} X$ and $B_2^{1/2} X$ is a linear manifold dense in X. If we take $a \in B_1^{1/2} X - B_2^{1/2} X$, then $\mu_a^1 \sim \mu^1$, $\mu_a^2 \perp \mu^1$, $\mu_a^2 \perp \mu^2$, $\mu_a^2 \perp \mu_a$ and $\mu_a^2 \perp \mu_1$ so that

$$\frac{d\mu_a}{d\mu}(x) = \frac{d\mu_a^1}{d\mu^1}(x) \quad \text{and} \quad \int \frac{d\mu_a}{d\mu}(x)\, \mu(dx) = \int \frac{d\mu_a^1}{d\mu^1}(x) \cdot \frac{1}{2} \mu^1(dx) = \frac{1}{2} \,.$$

Hence, μ will not be equivalent to a product measure, but M_μ will contain a linear manifold dense in X.

§ 21. Differentiation of Measures w.r.t. a Direction

We will say that a measure μ is *differentiable w.r.t. the direction* a (or *in the direction* a) if for any $f \in C_X$ (C_X is the set of all bounded continuous numerical-valued functions on X), the limit

$$\lim_{\lambda \downarrow 0} \frac{1}{\lambda} \left[\int f(x)\, \mu_{\lambda a}(dx) - \int f(x)\, \mu(dx)\right] = \lim_{\lambda \downarrow 0} \int \frac{f(x + \lambda a) - f(x)}{\lambda}\, \mu(dx) \,. \tag{1}$$

exists. We will denote this limit by $l(f)$. Since $l(f)$ is the limit of an everywhere convergent sequence of linear functionals on the Banach space C_X,

$$l(f) = \lim_{n\to\infty} l_n(f) \,,$$

where $l_\lambda(f) = \int \frac{1}{\lambda} [f(x + \lambda a) - f(x)] \mu(dx)$, $l(f)$ is also a linear functional on C_X; in particular, it follows from this that there exists a constant α such that

$$|l(f)| \leq \alpha \|f\| \,,$$

where $\|f\| = \sup_x |f(x)|$. In addition, there exists an α such that for all $\lambda > 0$

$$|l_\lambda(f)| \leq \alpha \|f\| \,. \tag{2}$$

This follows from the well-known theorem of Banach on the limit of linear functionals in a Banach space.

Theorem 1. *If the limit* (1) *is defined for all* $f \in C_X$ *then there exists on* $\mathfrak{B}$ *a countably-additive set function* ν *of bounded variation for which*

$$l(f) = \int f(x) \, \nu(dx) \,. \tag{3}$$

Proof. 1. Let $\nu_\lambda = \frac{1}{\lambda} (\mu_{\lambda a} - \mu)$. To prove (3) it is sufficient to show that for any $\varepsilon > 0$ we can find a compact $K_\varepsilon \subset X$ for which the variation of the measure ν_λ on $X - K_\varepsilon$ will be smaller than ε. In this case

$$l_\lambda(f) = l_\lambda(K_\varepsilon, f) + O\left(\varepsilon \sup_{x \notin K_\varepsilon} |f(x)|\right),$$

where $l_\lambda(K_\varepsilon, f) = \int_{K_\varepsilon} f(x) \, \nu_\lambda(dx)$. $l_\lambda(K_\varepsilon, l)$ is a linear functional on the separable Banach space C_{K_ε} of continuous functions defined on K_ε. Each linear functional on C_{K_ε} has the form (3). Moreover, an arbitrary bounded set of linear functionals on C_{K_ε} is weakly compact. (C_{K_ε} is separable.) Hence, for any ε the representation

$$l(f) = \int_{K_\varepsilon} f(x) \, \nu(\varepsilon, dx) + O\left(\varepsilon \sup_{x \notin K_\varepsilon} |f(x)|\right), \tag{4}$$

is valid, where $\nu(\varepsilon, \cdot)$ is a countably additive set function on K_ε. It is possible to choose K_ε so that $K_{\varepsilon_1} \supset K_{\varepsilon_2}$ for $\varepsilon_1 < \varepsilon_2$. Then $\nu(\varepsilon_2, \cdot)$ and $\nu(\varepsilon_1, \cdot)$ coincide on K_{ε_2}. Setting for all $B \in \mathfrak{B}$

$$\nu(B) = \lim_{\varepsilon \to 0} \nu(\varepsilon, B \cap K_\varepsilon) \,.$$

and going to the limit for $\varepsilon \to 0$ in (4), we obtain (3).

2. Let B be some set. Put $B^{(\delta)} = \bigcup_{x \in B} S_\delta(x)$, where $S_\delta(x)$ is an open sphere of radius δ with center at x. In order to show that for each $\varepsilon > 0$ there exists a compact K such that $\bar{\nu}_\lambda (X - K) < \varepsilon$ for all $\lambda > 0$, where $\bar{\nu}_\lambda$ is the variation of the measure ν_λ, it sufficies to prove that for

each $\varepsilon > 0$ and $\delta > 0$ there exists a compact $K_1^{(\delta)}$ for which $\bar{\nu}_\lambda (X - K_1^{(\delta)}) < \varepsilon$. Indeed, if this is the case, then choosing a sequence $\delta_n \downarrow 0$, we can, for all n, construct a compact $K_n^{(\delta_n)}$ for which

$$\bar{\nu}_\lambda (X - K_n^{(\delta_n)}) < \varepsilon/2^n .$$

Then, $K = \bigcup\limits_{n=1}^{\infty} K_n^{(\delta_n)}$ will be a compact (since $K \subset K_n^{(\delta_n)}$ and $K_n^{(\delta_n)}$ is a compact, K contains a finite $2\delta_n$-net). We have

$$\bar{\nu}_\lambda (X - K) \leq \sum_{n=1}^{\infty} \bar{\nu}_\lambda (X - K_n^{(\delta_n)}) \leq \sum_{n=1}^{\infty} \frac{\varepsilon}{2^n} = \varepsilon .$$

3. We assume that for some $\varepsilon > 0$ and $\delta > 0$ it is not possible to find a compact K such that $\bar{\nu}_\lambda (X - K) < \varepsilon$ for all $\lambda > 0$. Then one can find a sequence of compacts K_n and a sequence λ_n such that

$$\bar{\nu}_{\lambda_n}(K_n) > \frac{\varepsilon}{2}, \qquad \bar{\nu}_{\lambda_n}\left(X - \bigcup_{j=1}^{n} K_j\right) \leq \frac{\varepsilon}{8}$$

and $K_n \subset X - \bigcup\limits_{j=1}^{n-1} K_j^{(\delta)}$. Hence, the sets $K_n^{(\delta/2)}$ are pairwise disjoint. For each n one can find a continuous function $f_n(x)$ equal to zero on $X - K_n^{(\delta/2)}$ and such that

$$\int f_n(x)\, \nu_{\lambda_n}(dx) \geq \varepsilon/2 \qquad \text{and} \qquad \|f_n\| \leq 1 .$$

Since the sets $K_n^{(\delta/2)}$ are disjoint for different n, for an arbitrary set of n_k, the series $\sum\limits_k f_{n_k}(x)$ converges and represents a continuous function with norm no greater than 1. For each prime number p we construct the function $g_p(x)$:

$$g_p(x) = \sum_j f_{p^{n_j}}(x) ,$$

where the finite or infinite set $\{n_j\}$ is chosen in such a way that for all k

$$\left|\int \sum_j f_{p^{n_j}}(x)\, \tilde{\nu}_p(dx)\right| > \frac{\varepsilon}{\delta}, \tag{5}$$

and $\bar{\nu}_n = \nu_{\lambda_n}$. We will show how to obtain (5). Choose $n_1 = 1$ and if $n_1, n_2, \ldots, n_k$ have been chosen, set

$$n_{k+1} = \inf \left\{m\colon m > n_k \text{ and} \left|\int \left(\sum_{j=1}^{k} f_{p^{n_j}}(x) + f_{pm}(x)\right) \tilde{\nu}_{pm}(dx)\right| > \frac{\varepsilon}{4}\right\}. \tag{6}$$

If the set in curled brackets is empty, then $\{n_1, \ldots, n_k\}$ will be the sought-for set of indices. From the definition of n_{k+1}, we have for $n_k < m < n_{k+1}$

$$\left|\int \left(\sum_{j=1}^{k} f_{p^{n_j}}(x) + f_{p^m}(x)\right) \tilde{\nu}_{p^m}(dx)\right| < \frac{\varepsilon}{4} .$$

Since

$$\int f_{pm}(x)\, \tilde{\nu}_{pm}(dx) \geq \frac{\varepsilon}{2},$$

$$\left|\int \sum_{j=1}^{k} f_{p^{n_j}}(x)\, \tilde{\nu}_{pm}(dx)\right| \geq \frac{\varepsilon}{4} \tag{7}$$

for $n_k < m < n_{k+1}$, and if n_{k+1} does not exist, then for all $m > n_k$. Thus,

$$\left|\int \sum_{j} f_{p^{n_j}}(x)\, \tilde{\nu}_{pm}(dx)\right| \geq \left|\int \sum_{n_j \leq m} f_{p^{n_j}}(x)\, \tilde{\nu}_{pm}(dx)\right| -$$

$$- \left|\int \sum_{n_j > m} f_{p^{n_j}}(x)\, \tilde{\nu}_{pm}(dx)\right| \geq \frac{\varepsilon}{4} - \nu_{\lambda_{pm}}\left(X - \bigcup_{i=1}^{pm} K_i\right) \geq \frac{\varepsilon}{8}.$$

The set $\{n_k\}$ has thus been constructed and with it also the functions $g_p(x)$. Without loss of generality we can assume that λ_n has a limit inasmuch as $\tilde{\nu}_{\lambda_n}(X) \leq \frac{2}{\lambda_n}$ so that from the assumption $\tilde{\nu}_{\lambda_n}(X) > \varepsilon$, it follows that $\lambda_n < \frac{2}{\varepsilon}$. Since

$$\lim_{n\to\infty} \int g_p(x)\, \tilde{\nu}_{\lambda_n}(dx) = \lim_{m\to\infty} \int g_p(x)\, \tilde{\nu}_{pm}(dx),$$

$$\left|\lim_{n\to\infty} \int g_p(x)\, \nu_{\lambda_n}(dx)\right| \geq \frac{\varepsilon}{8}.$$

We choose ε_p with $|\varepsilon_p| = 1$ so that

$$\lim_{n\to\infty} \int \varepsilon_p \cdot g_p(x) \cdot \nu_{\lambda_n}(dx) \geq \frac{\varepsilon}{8}.$$

Then for each N

$$\lim_{n\to\infty} \int \sum_{p=1}^{N} \varepsilon_p\, g_p(x)\, \nu_{\lambda_n}(dx) \geq N \cdot \frac{\varepsilon}{8}. \tag{8}$$

But $\left\| \sum_{p=1}^{N} \varepsilon_p\, g_p(x) \right\| \leq 1$ for all N since the functions $g_p(x)$ do not exceed 1 in norm and for different p cannot all be different from zero. From (2) we obtain

$$N \cdot \frac{\varepsilon}{8} \leq \lim_{n\to\infty} \int \sum_{p=1}^{N} \varepsilon_p \cdot g_p(x)\, \nu_{\lambda_n}(dx) \leq \alpha,$$

(α is taken from (2)), which is impossible for $N > \frac{8\alpha}{\varepsilon}$. This contradiction implies the validity of the claim in the theorem. □

We have thus established the existence of a countably additive function ν such that for all $f \in C_X$

$$\lim_{\lambda \downarrow 0} \int \frac{1}{\lambda} [f(x + \lambda a) - f(x)]\, \mu(dx) = \int f(x)\, \nu(dx), \tag{9}$$

if μ is differentiable in the direction a. This measure ν will be called the *derivative of μ in the direction a* and is written as

$$\nu = d_a\mu .$$

Theorem 2. *If $d_a\mu$ exists for the measure μ, then for all $\alpha > 0$ and each bounded $\mathfrak{B}$-measurable function f*

$$\int [f(x+\alpha a) - f(x)]\,\mu(dx) = \int \left[\int_0^\alpha f(x+\lambda a)\,d\lambda\right] d_a\mu(dx) \qquad (10)$$

and the function $d_a\mu$ uniquely determines the measure μ.

Proof. It suffices to establish the relation (10) for $f \in C_X$. In this case

$$\int \left[\int_0^\alpha f(x+\lambda a)\,d\lambda\right] d_a\mu(dx) = \int_0^\alpha [f(x+\lambda a)\,d_a\mu(dx)]\,d\lambda$$

and the function $\int f(x+\lambda a)\,\nu(dx)$ is continuous in λ. Hence,

$$\frac{d}{d\alpha}\int_0^\alpha [\int f(x+\lambda a)\,d_a\mu(dx)]\,d\lambda = \int f(x+\lambda a)\,d_a\mu(dx)$$

exists. The function $\int f(x+\alpha a)\,\mu(dx)$ is continuous from the right w.r.t. α due to the existence of $d_a\mu$ and if $\frac{d^+}{d\alpha}$ denotes the right derivative, then

$$\frac{d^+}{d\alpha}\int f(x+\alpha a)\,\mu(dx) = \int f(x+\lambda a)\,d_a\mu(dx) .$$

This means that the function

$$\xi(\alpha) = \int [f(x+\alpha a) - f(x)]\,\mu(dx) - \int \left[\int_0^\alpha f(x+\lambda a)\,d\lambda\right] d_a\mu(dx)$$

is continuous w.r.t. α and has a continuous right-derivative w.r.t. α:

$$\frac{d^+}{d\alpha}\xi(\alpha) = 0 \qquad \text{for } \alpha > 0 ;$$

moreover, $\xi(0) = 0$. Therefore, $\xi(\alpha) = 0$ for $\alpha > 0$ and (10) is proved. In order to demonstrate the possibility of defining the measure μ by means of $d_a\mu$, we note that for any function $f \in C_X$ for which $f(x) \to 0$ when $|x| \to \infty$, one has

$$\lim_{\alpha\to\infty} \int f(x+\alpha a)\,\mu(dx) = 0 .$$

Hence, for such functions

$$\int f(x)\,\mu(dx) = -\lim_{\alpha\to\infty} \int \left[\int_0^\alpha f(x+\lambda a)\,d\lambda\right] d_a\mu(dx) . \qquad (11)$$

It is clear that if we know the value of the integral $\int f(x)\,\mu(dx)$ for $f \in C_X$ satisfying the condition $f(x) \to 0$ for $|x| \to \infty$, then we can define this integral for all $f \in C_X$. By the same token it is uniquely defined. □

We will be interested in the connection between admissible directions and derivatives w.r.t. a direction. We note some simple properties of derivatives w.r.t. a direction.

1. If μ is differentiable in the direction a, then for all $b \in X$ the measure μ_b will also be differentiable in the direction a, and

$$d_a(\mu_b) = (d_a\mu)_b\,, \tag{12}$$

where $(d_a\mu)_b$ denotes the countably-additive set function obtained from the function $d_a\mu$ by a shift of b units:

$$\int f(x)\,(d_a\mu)_b\,(dx) = \int f(x+b)\,d_a\mu(dx)\,.$$

2. Let a be an admissible direction for the measure $\nu = d_a\mu$. Then μ is absolutely continuous w.r.t. ν. In fact, starting from (10) and writing $\frac{d\nu_{\lambda a}}{d\nu}(x) = \varrho_\nu(\lambda a, x)$, we have

$$\int f(x)\,\mu_{\alpha a}(dx) = \int f(x)\,\mu(dx) + \int f(x)\left[\int_0^\alpha \varrho_\nu(\lambda a, x)\,d\lambda\right]\nu(dx)\,. \tag{13}$$

Let A be a bounded set for which $\bar{\nu}(A) = 0$, where $\bar{\nu}$ is the variation of the measure ν. Setting $f(x) = \chi_A(x)$ in (13) we get

$$\mu_{\alpha a}(A) = \mu(A)$$

for all $\alpha > 0$. Letting $\alpha \to \infty$, we see that $\mu(A) = 0$. Hence, $\mu \ll \nu$ and $\mu_{\alpha a} \ll \nu$.

3. Let $\varrho(x)$ be the density of μ w.r.t. ν and $\varrho_\alpha(x)$ that of $\mu_{\alpha a}$ w.r.t. ν. Then it follows from (13) that

$$\varrho_\alpha(x) = \varrho(x) + \int_0^\alpha \varrho_\nu(\lambda a, x)\,d\lambda \qquad (\operatorname{mod} \bar{\nu})\,. \tag{14}$$

Thus, for almost all x $(\operatorname{mod} \bar{\nu})$ $\frac{d}{d\alpha}\varrho_\alpha(x) = \varrho_\nu(\alpha a, x)$ exists for almost all $\alpha > 0$ (w.r.t. Lebesgue measure). Note that from

$$\int f(x)\,\mu_{\alpha a}(dx) = \int f(x+\alpha a)\,\varrho(x)\,\nu(dx) = \int f(x)\,\varrho(x-\alpha a)\,\varrho_\nu(\alpha a, dx)$$

it follows that

$$\varrho_\alpha(x) = \varrho(x-\alpha a)\,\varrho_\nu(\alpha a, x) \qquad (\operatorname{mod} \bar{\nu})\,.$$

Hence, (14) is equivalent to the relation

$$\varrho(x-\alpha a)\,\varrho_\nu(\alpha a, x) = \varrho(x) + \int_0^\alpha \varrho_\nu(\lambda a, x)\,d\lambda\,, \tag{15}$$

whence we get

$$\frac{d}{d\alpha}\left(\varrho(x-\alpha a)\,\varrho_\nu(\alpha a, x)\right) = \varrho_\nu(\alpha a, x)\,. \tag{16}$$

If for some x $\varrho(x-\alpha a)$ does not vanish (for sufficiently small α), then (16) can be considered as a differential equation for ϱ_ν. Solving this equation we get

$$\varrho(x-\alpha a)\,\varrho_\nu(\alpha a,x) = \varrho(x)\exp\left\{\int\limits_0^\alpha \frac{d\lambda}{\varrho(x-\lambda a)}\right\}. \tag{17}$$

4. Assume that (17) holds for almost all x (mod μ) for sufficiently small α. Then III § 19 implies that $\mu_{\alpha a} \ll \mu$ for all small enough α and

$$\varrho_\mu(\alpha a,x) = \frac{\varrho(x-\alpha a)}{\varrho(x)}\,\varrho_\nu(\alpha a, x) = \exp\left\{\int\limits_0^\alpha \frac{d\lambda}{\varrho(x-\lambda a)}\right\}. \tag{18}$$

If (2) § 19 is invoked, then we can show that (18) holds for all $\alpha > 0$.

5. If the measure μ is differentiable in the directions a and b, then it is also differentiable in the direction $a+b$, and

$$d_{a+b}\mu = d_a\mu + d_b\mu\,.$$

Indeed, using (10) we can write for $f \in C_X$

$$\begin{aligned}
&\int [f(x+\alpha a+\alpha b) - f(x)]\,\mu(dx) = \\
&= \int [f(x+\alpha a+\alpha b) - f(x+\alpha a)]\,\mu(dx) + \\
&\quad + \int [f(x+\alpha a) - f(x)]\,\mu(dx) = \\
&= \int\int\limits_0^\alpha f(x+\alpha a+\lambda b)\,d\lambda\, d_b\mu(dx) + \int\int\limits_0^\alpha f(x+\lambda a)\,d\lambda\, d_a\mu(dx)\,.
\end{aligned}$$

Dividing this relation by α and letting $\alpha \downarrow 0$, we obtain the claimed result since

$$\frac{1}{\alpha}\int\limits_0^\alpha f(x+\alpha a+\lambda b)\,d\lambda \qquad \text{and} \qquad \frac{1}{\alpha}\int\limits_0^\alpha f(x+\lambda a)\,d\lambda$$

are bounded and converge to $f(x)$ for $\alpha \downarrow 0$.

6. If the measure μ is differentiable in the direction a, then for all $\xi \in (-\infty, \infty)$ it will also be differentiable in the direction ξa, and

$$d_{\xi a}\mu = \xi\, d_a\mu\,.$$

For $\xi > 0$ we have from (10)

$$\begin{aligned}
\int [f(x+\alpha\xi a) - f(x)]\,\mu(dx) &= \int\int\limits_0^{\alpha\xi} f(x+\lambda a)\,d\lambda\, d_a\mu(dx) = \\
&= \xi\int\int\limits_0^\alpha f(x+\lambda\xi a)\,d\lambda\, d_a\mu(dx)\,.
\end{aligned}$$

We now show that $d_{-a}\mu = -d_a\mu$. To this end, we substitute the function $f(x - \alpha a)$ into (10) and get

$$\int [f(x) - f(x - \alpha a)]\, \mu(dx) = \int \int_0^\alpha f(x - \alpha a + \lambda a)\, d\lambda\, d_a\mu(dx) =$$

$$= \int \int_{-\alpha}^{\alpha} f(x - \lambda a)\, d\lambda\, d_a\mu(dx)\,.$$

Dividing by α and letting $\alpha \downarrow 0$ we prove the assertion.

Definition. If the measure μ is differentiable in the direction a and $d_a\mu \ll \mu$, then the density of the countably-additive function $d_a\mu$ w.r.t. μ is called the *logarithmic derivative* of the measure μ in the direction a. We will denote it by $l_\mu(a, x)$:

$$l_\mu(a, x) = \frac{d(d_a\mu)}{d\mu}(x)\,. \tag{19}$$

Relation (10) implies that when $l_\mu(a, x)$ exists, then for all $\alpha > 0$

$$\int [f(x + \alpha a) - f(x)]\, \mu(dx) = \int \left[\int_0^\alpha f(x + \lambda a)\, d\lambda\right] l_\mu(a, x)\, \mu(dx)\,. \tag{20}$$

Using Properties 5 and 6 we see that the set D_l of all a for which the function $l_\mu(a, x)$ is defined is a linear manifold and for $a, b \in D_l$

$$l_\mu(\lambda_1 a + \lambda_2 b, x) = \lambda_1 l_\mu(a, x) + \lambda_2 l_\mu(b, x) \qquad (\bmod \mu)\,. \tag{21}$$

Let X be a finite-dimensional space and assume the measure μ has a positive density w.r.t. Lebesgue measure. Assume that the measure μ has a logarithmic derivative in the direction a. Denote by $m(dx)$ Lebesgue measure in X and set $p(x) = \frac{d\mu}{dm}(x)$. Then from (20)

$$\int f(x)\, [p(x - \alpha a) - p(x)]\, m(dx) =$$

$$= \int f(x) \int_0^\alpha l_\mu(a, x - \lambda a)\, p(x - \lambda a)\, d\lambda\, m(dx)\,.$$

Hence, for almost all x (w.r.t. Lebesgue measure)

$$p(x - \alpha a) - p(x) = \int_0^\alpha l_\mu(a, x - \lambda a)\, p(x - \lambda a)\, d\lambda$$

so that the derivative w.r.t. α of $p(x - \alpha a)$ exists:

$$\frac{d}{d\alpha} p(x - \alpha a) = l_\mu(a, x - \alpha a)\, p(x - \alpha a)\,.$$

From the latter we obtain

$$l_\mu(a, x) = \frac{1}{p(x)} \cdot \frac{d}{d\alpha} p(x - \alpha a)\Big|_{\alpha=0} = \frac{d}{d\alpha} \log p(x - \alpha a)\Big|_{\alpha=0}\,.$$

From this it follows that in the finite-dimensional case, $-l_\mu(a, x)$ is the logarithmic derivative of the density of μ w.r.t. Lebesgue measure in the direction a.

Let L be a finite-dimensional subspace of the Hilbert space X and μ_L the projection of the measure μ on L. If μ is differentiable in the direction a, then μ_L will be differentiable in the direction $a' = P_L a$, where P_L is projection on L. In this connection, $d_{a'}\mu_L$ will coincide with $(d_a\mu)_L$, the projection of the countably-additive function $d_a\mu$ on L. Indeed, if $\varphi(x)$ is a bounded continuous function on L and $f \in C_X$ is such that $f(x) = \varphi(P_L x)$, then applying (10) to f, we get

$$\int [\varphi(x + \alpha a') - \varphi(x)]\, \mu_L(dx) = \int [f(x + \alpha a) - f(x)]\, \mu(dx) =$$

$$= \int \int_0^\alpha f(x + \lambda a)\, d\lambda\, d_a\mu(dx) = \int \int_0^\alpha \varphi(P_L x + \lambda a')\, d\lambda\, d_a\mu(dx) =$$

$$= \int \int_0^\alpha \varphi(x + \lambda a')\, d\lambda (d_a\mu)_L\, (dx)\,.$$

If μ has a logarithmic derivative in the direction a, i.e., $d_a\mu \ll \mu$, then $(d_a\mu)_L \ll \mu_L$. Hence, μ_L will also have a logarithmic derivative in the direction a', and by (19) § 15

$$\frac{d(d_{a'}\mu_L)}{d\mu_L}(x) = \int l_\mu(a, x')\, \mu(dx', \mathfrak{B}^L / P_L^{-1} x)\,, \tag{22}$$

where $\mu(\cdot, \mathfrak{B}^L/x)$ is the conditional measure for μ w.r.t. the σ-algebra $\mathfrak{B}^L$ of sets of the form $P_L^{-1}(A)$ with A a Borel set in L.

§ 22. An Admissibility Condition for Shifts

In this section we will consider a special class of measures μ satisfying the following two conditions:

I. It is possible to find a sequence of finite-dimensional subspaces L_n for which $L_n \subset L_{n+1}$ and $\bigcup L_n$ is dense in X and for each n the projection of the measure μ on L_n has a density $p_n(x)$ w.r.t. Lebesgue measure on L_n.

II. The density $p_n(x)$ is almost everywhere positive w.r.t. Lebesgue measure on L_n and for a given a there exists a measurable function $l_n(x)$ on L_n such that $\int |l_n(x)|\, p_n(x)\, dx < \infty$ and for all α

$$p_n(x - \alpha a_n) - p(x) = \int_0^\alpha l_n(x - \lambda a_n)\, p_n(x - \lambda a)\, d\lambda\,, \tag{1}$$

where $a_n = P_n a$ and P_n is the projection operator on the subspace L_n.

We will derive a condition below which will ensure that the vector a mentioned in II will be an admissible direction for the measure μ.

Let us discuss the nature of Conditions I and II. From the results of § 20 it follows that Condition I and the positivity of the density $p_n(x)$ are necessary to ensure that $L_n \subset M_\mu$ for all n. Condition II holds (see § 21) if the measure μ_{L_n} is logarithmically differentiable in the direction a_n. This condition is necessary for logarithmic differentiability in the direction a. It will turn out that logarithmic differentiability of the measure μ in the direction a is closely connected with the behavior of $l_n(x)$ for $n \to \infty$.

Theorem 1. *The sequence of functions $l_n(P_n x)$ is a martingale. The logarithmic derivative of the measure μ in the direction a exists iff the sequence $l_n(P_n x)$ is uniformly integrable and then we have*

$$l_\mu(a, x) = \frac{d(d_a\mu)}{d\mu}(x) = \lim_{n\to\infty} l_n(P_n x) \qquad (\text{mod } \mu) \tag{2}$$

Proof. From (22) § 21 it follows for $n < m$ that

$$\frac{d(d_{a_n}\mu)_{L_n}}{d\mu_{L_n}}(P_{L_n} x) = \int \frac{d(d_{a_m}\mu)_{L_m}}{d\mu_{L_m}}(x')\, \mu_{L_m}(dx', \mathfrak{B}_{L_m}^{L_n}/x), \tag{3}$$

where μ_{L_n} is the projection of μ on L_n, $\mu_{L_m}(\cdot, \mathfrak{B}_{L_m}^{L_n}/x)$ is the conditional measure for μ_{L_m} w.r.t. the σ-algebra $\mathfrak{B}_{L_m}^{L_n}$ of sets in L_m of the form $L_m \cap \{x\colon P_{L_n} x \in A\}$ and $A \in \mathfrak{B}_{L_n}$, the σ-algebra of Borel sets in L_n. Taking into account the fact that μ_{L_m} is the projection of μ on L_m, we find that

$$\mu_{L_m}(A, \mathfrak{B}_{L_m}^{L_n}/x) = \mu(P_{L_m}^{-1} A, \mathfrak{B}^{L_m}/P_{L_m}^{-1} x)\,.$$

Thus, (3) can be written as

$$\hat{l}_n(x) = \int \hat{l}_m(x')\, \mu(dx', \mathfrak{B}^{L_n}/x), \hat{l}_n(x) = l_n(P_n x)\,. \tag{4}$$

The function $\hat{l}_n(x)$ is $\mathfrak{B}^{L_n}$-measurable. Hence, for any bounded $\mathfrak{B}^{L_n}$-measurable function $\psi(x)$ we find from (10) § 13 that

$$\int \hat{l}_n(x)\, \psi(x)\, \mu(dx) = \int \int \hat{l}_m(x')\, \mu(dx', \mathfrak{B}^{L_n}/x)\, \psi(x)\, \mu(dx) =$$
$$= \int \hat{l}_m(x)\, \psi(x)\, \mu(dx)\,.$$

This proves that $\hat{l}_n(x)$ is a martingale.

If μ is logarithmically differentiable in the direction a, then (22) § 21 implies that

$$\hat{l}_n(x) = \int l_\mu(a, x')\, \mu(dx', \mathfrak{B}^{L_n}/x)\,. \tag{5}$$

Set $\gamma_N(\lambda) = 0$ for $\lambda < N$ and $\gamma_N(\lambda) = \lambda - N$ for $\lambda \geq N$. To demonstrate the uniform integrability it suffices to prove that

$$\lim_{N\to\infty} \sup_n \int \gamma_N(|\hat{l}_n(x)|)\, \mu(dx) = 0\,. \tag{6}$$

Since the function $\gamma_N(|x|)$ is convex downward, we get from Jensen's inequality

$$\gamma_N(|\hat{l}_n(x)|) \leq \int \gamma_N\left(|l_\mu(a, x')|\right) \mu(dx', \mathfrak{B}^{L_n}/x),$$

so that

$$\int \gamma_N\left(|\hat{l}_n(x)|\right) \mu(dx) \leq \int \int \gamma_N(|l_\mu(a, x')|) (dx', \mathfrak{B}^{L_n}/x) \mu(dx) = \\ = \int \gamma_N(|l_\mu(a, x')|) \mu(dx).$$

Since $\gamma_N(|l_\mu(a, x)|) \leq |l_\mu(a, x)|$, $\int |l_\mu(a, x)| \mu(dx) < \infty$ and $\gamma_N(|l_\mu(a, x)|) \leq \leq |l_\mu(a, x)|$ when $N \to \infty$ for μ-almost all x, we get

$$\lim_{N\to\infty} \int \gamma_N(|l_\mu(a, x')|) \mu(dx') \to 0.$$

From the last inequality we obtain (6).

Now let the sequence $\hat{l}_n(x)$ be uniformly integrable. Then, in particular, $\sup_n \int |\hat{l}_n(x)| \mu(dx) < \infty$, so that by Theorem 1, § 14, the limit

$$\lim_{n\to\infty} \hat{l}_n(x) = l(x)$$

exists μ-a.e. Let $\varphi(x)$ be an arbitrary continuous function on L_n. Multiplying (1) by $\varphi(x)$ and integrating w.r.t. Lebesgue measure in L_n, we get

$$\int [\varphi(x + \alpha a_n) - \varphi(x)] \mu_{L_n}(dx) = \int \int_0^\alpha \varphi(x + \lambda a_n) \, d\lambda \, l_n(x) \mu_{L_n}(dx).$$

For some m let $f(x) = \varphi(P_m x)$, where φ is continuous and bounded on L_m. We have for all $n > m$

$$\int [f(x + a) - f(x)] \mu(dx) = \int \int_0^\alpha f(x + \lambda a) \, d\lambda \, \hat{l}_n(x) \mu(dx). \tag{7}$$

The uniform integrability of $\hat{l}_n(x)$ allows the limit passage $n \to \infty$ in (7). Carrying this out, we find that for any function of the form $f(x) = \varphi(P_m x)$, where φ is continuous and bounded on L_m,

$$\int [f(x + a) - f(x)] \mu(dx) = \int \int_0^\alpha f(x + \lambda a) \, d\lambda \, l(x) \mu(dx). \tag{8}$$

Using a limit passage on f we convince ourselves of the validity of (8) for all $f \in C_X$. From (8) it follows that $d_a\mu$ exists and

$$l(x) = \frac{d(d_a\mu)}{d\mu}(x). \quad \square$$

Theorem 2. *Assume Conditions* I *and* II *hold and that for sufficiently small* $\delta > 0$

$$\sup_n \int e^{\delta|l_n(P_n x)|} \mu(dx) < \infty. \tag{9}$$

Then for all λ, $\lambda a \in M_\mu$ and

$$\varrho_\mu(\lambda a, x) = \exp\left\{\int_0^\lambda l_\mu(a, x - \alpha a)\, d\alpha\right\}, \tag{10}$$

where $l_\mu(a, x)$ is the logarithmic derivative of μ in the direction a.

Proof. We note immediately that the existence of $l_\mu(a, x)$ follows from Theorem 1 since (9) implies the uniform integrability of l_n. To show that $\lambda a \in M_\mu$, it is sufficient to prove the uniform integrability w.r.t. n of the sequence of functions $g_n(x) = p(P_n x - a_n)/p(P_n x)$, which are the densities for $x \in L_n$ of $(\mu_{L_n})_{\alpha_n}$ w.r.t. μ_{L_n}, where $(\mu_{L_n})_{a_n}$ are the μ_{L_n} shifted by a_n. To do this it sufficies to demonstrate the uniform boundedness w.r.t. n of the integrals

$$I_n(\lambda) = \int \frac{p_n(x - \lambda a_n)}{p_n(x)} \log\left(\frac{p_n(x - \lambda a_n)}{p_n(x)}\right) p_n(x)\, m_n(dx)\,,$$

where m_n is Lebesgue measure on L_n. Since

$$\log \frac{p_n(x - \lambda a_n)}{p_n(x)} = \int_0^\lambda l_n(x - \alpha a_n)\, d\alpha$$

for almost all $x \in L_n$ w.r.t. Lebesgue measure,

$$\begin{aligned} I_n(\lambda) &= \int p_n(x - \lambda a_n) \int_0^\lambda l_n(x - \alpha a_n)\, d\alpha\, m_n(dx) = \\ &= \int \int_0^\lambda l_n(x + \alpha a_n)\, d\alpha\, p_n(x)\, m_n(dx) = \\ &= \int l_n(x) \int_0^\lambda p_n(x - \alpha a_n)\, d\alpha\, m_n(dx)\,. \end{aligned}$$

We will assume that $\lambda > 0$ (we can treat $\lambda < 0$ by changing the sign of a). It follows from Young's inequality (see Hardy, Littlewood and Pólya [1]) that for $\alpha > 0$, $\beta > 0$ and $\delta > 0$

$$\alpha\beta \le \frac{1}{\delta}\alpha \ln(1 + \alpha) + \frac{e^{\delta\beta} - 1}{\delta}.$$

Hence,

$$|I_n(\lambda)| \le \int_0^\lambda \int |l_n(x)|\, p_n(x - \alpha a)\, d\alpha\, m_n(dx) \le$$

$$\le \int_0^\lambda \int \left(\frac{e^{\delta|l_n(x)|} - 1}{\delta} + \frac{1}{\delta} \cdot \frac{p_n(x - \alpha a_n)}{p_n(x)} \log\left(1 + \frac{p_n(x - \alpha a_n)}{p_n(x)}\right)\right) p_n(x)\, m_n(dx)\, d\alpha \le$$

$$\le \lambda \frac{1}{\delta} \int e^{\delta|l_n(P_n x)|} \mu(dx) +$$

$$+ \frac{1}{\delta} \int_0^\lambda \int \left[\frac{p_n(x - \alpha a_n)}{p_n(x)} \log \frac{p_n(x - \alpha a_n)}{p_n(x)} + 1\right] p_n(x)\, m_n(dx)\, d\alpha =$$

$$= \frac{1}{\delta}\left(\lambda \int e^{\delta|l_n(p_n x)|} \mu(dx) + \int_0^\lambda I_n(\alpha)\, d\alpha + \lambda\right).$$

(We have used the inequality $\alpha \log(1+\alpha) \leq \alpha \log \alpha + 1$.) Then for all $\lambda > 0$

$$|I_n(\lambda)| \leq \frac{\gamma}{\delta}\lambda + \frac{1}{\delta}\int_0^\lambda |I_n(\alpha)|\, d\alpha\,, \tag{11}$$

where $\gamma = 1 + \sup\limits_n \int e^{\delta|l_n(x)|}\,\mu(dx)$. Using iteration we get from (11)

$$|I_n(\lambda)| \leq \frac{\gamma}{\delta}\,(e^{\lambda/\delta} - 1)\,. \tag{12}$$

The uniform boundedness of $I_n(\lambda)$ is established and we have shown that $\lambda a \in M_\mu$. We turn to the derivation of (10). Using the already proved existence of $\varrho_\mu(\lambda a, x)$ and the equalities

$$\int |l_\mu(a, x - \lambda a)|\, \varrho_\mu(\lambda a, x)\, \mu(dx) = \int |l_\mu(a, x)|\, \mu(dx)\,,$$

and

$$\int_0^\lambda \int |l_\mu(a, x - \alpha a)|\, \varrho_\mu(\alpha a, x)\, \mu(dx)\, d\alpha = \lambda \int |l_\mu(a, x)|\, \mu(dx)\,,$$

we find that for μ-almost all x

$$\int_0^\lambda |l_\mu(a, x - \alpha a)|\, \varrho_\mu(\alpha a, x)\, \mu(dx)$$

exists. From (20) § 21 we have

$$\int f(x)\,[\varrho_\mu(\lambda a, x) - 1]\,\mu(dx) = \int f(x) \int_0^\lambda l_\mu(a, x - \alpha a)\, \varrho_\mu(\alpha a, x)\, d\alpha\, \mu(dx)$$

so that

$$\varrho_\mu(\lambda a, x) - 1 = \int_0^\lambda l_\mu(a, x - \alpha a)\, \varrho_\mu(\alpha a, x)\, d\alpha \qquad (\mathrm{mod}\, \mu)\,. \tag{13}$$

If for a given x and some $\lambda_1 > 0$ the integral

$$\int_0^{\lambda_1} |l_\mu(a, x - \alpha a)|\, \varrho_\mu(\alpha a, x)\, d\alpha\,,$$

exists, then one can find an $\varepsilon > 0$ such that for $0 \leq \lambda \leq \varepsilon$,

$$\int_0^\lambda |l_\mu(a, x - \alpha a)|\, \varrho_\mu(\alpha a, x)\, dx < \frac{1}{2}\,,$$

and then $\varrho_\mu(\alpha a, x) > \frac{1}{2}$, which implies the convergence of the integral

$$\int_0^\lambda |l_\mu(a, x - \alpha a)|\, d\alpha\,.$$

From (13) we get by iteration

$$\varrho_\mu(\lambda a, x) = \exp\left\{\int_0^\lambda l_\mu(a, x - \alpha a)\, d\alpha\right\}$$

for all x such that $\int_0^\lambda |l_\mu(a, x - \alpha a)| \, d\alpha < \infty$. In order to see that the latter inequality holds for μ-almost all x we consider the integral (assume $\lambda > 0$)

$$I_\lambda = \int \int_0^\lambda |l_\mu(a, x - \alpha a)| \, \mu_{-\lambda a}(dx) = \int |l_\mu(a, x)| \int_0^\lambda \varrho_\mu(\alpha a, x) \, d\alpha \, \mu(dx) .$$

Applying Young's inequality again, we get

$$\begin{aligned} I_\lambda &\leq \frac{\lambda}{\delta} \int (e^{\delta |l_\mu(a,x)|} - 1) \, \mu(dx) + \\ &+ \frac{1}{\delta} \int_0^\lambda \int [(\log \varrho_\mu(\alpha a, x)) \, \varrho_\mu(\alpha a, x) + 1] \, \mu(dx) \, d\alpha \leq \\ &\leq \frac{1}{\delta} \int e^{\delta |l_\mu(a,x)|} \, \mu(dx) + \\ &+ \frac{1}{\delta} \int_0^\lambda \int (\varrho_\mu(\alpha a, x) \log \varrho_\mu(\alpha a, x) + 1) \, \mu(dx) \, d\alpha . \end{aligned}$$

Since $l_\mu(a, x) = \lim_{n\to\infty} l_n(x) \pmod{\mu}$,

$$\int e^{\delta |l_\mu(a,x)|} \, \mu(dx) \leq \sup_n \int e^{\delta l_n(P_n x)} \, \mu(dx) ,$$

and

$$\varrho_\mu(\alpha a, x) = \lim_{n\to\infty} \frac{p_n(P_n x - \alpha P_n a)}{p_n(P_n x)} .$$

Hence, by Fatou's Lemma

$$\int_0^\lambda \int [\varrho_\mu(\alpha a, x) \log \varrho_\mu(\alpha a, x) + 1)] \, d\alpha \, \mu(dx) \leq \varliminf_{n\to\infty} \int_0^\lambda [I_n(\alpha) + 1] \, d\alpha .$$

The boundedness of I_λ now follows from (9) and (12). □

Remark 1. Since (9) implies the existence of $l_\mu(a, x)$ and the inequality

$$\int e^{\delta |l_\mu(a,x)|} \, \mu(dx) < \infty , \tag{14}$$

and (14) along with Jensen's inequality and (5) imply

$$\begin{aligned} \int e^{\delta |l_n(P_n x)|} \, \mu(dx) &= \int \exp \{\delta \, |\int l_\mu(a, x') \, \mu(dx', \mathfrak{B}^{L_n}/x)|\} \, \mu(dx) \leq \\ &\leq \int \int \exp \{\delta \, |l_\mu(a, x')|\} \, \mu(dx', \mathfrak{B}^{L_n}/x) \, \mu(dx) = \\ &= \int e^{\delta |l_\mu(a,x)|} \, \mu(dx) , \end{aligned}$$

we find that (9) is equivalent to (14). The advantage of (9) is that finite-dimensional distributions are used for its verification.

Remark 2. Since the assumptions of the theorem can be formulated in terms of the logarithmic derivative, it is natural to assume that

Conditions I and II, quoted at the begining of the section, are superfluous when $l_\mu(a, x)$ exists and (14) holds. This is actually the case. To show this, we consider in place of μ the measure $\bar{\mu}^\varepsilon$ equal to the convolution of μ and ν_ε, where ν_ε is a Gaussian measure with mean 0 and correlation operator εB and B is some nuclear operator. If $L_n \subset B^{1/2} X$, then the measure $\bar{\mu}^\varepsilon$ satisfies Conditions I and II of this section. Moreover,

$$l_{\bar{\mu}^\varepsilon}(a, x) = \int l_\mu(a, x') \, \mu \times \nu_\varepsilon \left(dx_1 \times dx_2, (\mathfrak{B} \times \mathfrak{B})^+/\sigma^{-1}(x)\right) \tag{15}$$

(for the notation see (24) § 15). Formula (15) follows from (24) and the fact that for any $f \in C_X$

$$\int [f(x + y + \lambda a) - f(x + y)] \, \mu(dx) \, \nu_\varepsilon(dx) =$$
$$= \int \int_0^\lambda f(x + y + \alpha a) \, d\alpha \, l_\mu(a, x) \, \mu(dx) \, \nu_\varepsilon(dy) \, .$$

From (15) we get

$$\int \exp\{\delta \, |l_{\bar{\mu}^\varepsilon}(a, x)|\} \, \bar{\mu}^\varepsilon(dx) \leq$$
$$\leq \int \exp\{\delta \int |l_\mu(a, x_1)| \, \mu \times \nu_\varepsilon \, (dx_1 \times dx_2, (\mathfrak{B} \times \mathfrak{B})^+/x_1 + x_2)\} \, \mu(dx_1) \, \nu_\varepsilon(dx_2) \leq$$
$$\leq \int \exp\{\delta \, |l_\mu(a, x_1)|\} \, \mu(dx_1) \, \nu_\varepsilon(dx_2) =$$
$$= \int \exp\{\delta \, |l_\mu(a, x_1)|\} \, \mu(dx) \, . \tag{16}$$

This means that the measure $\bar{\mu}^\varepsilon$ also satisfies (14). Let $p_n^\varepsilon(x)$ be the density of $\bar{\mu}^\varepsilon$ w.r.t. Lebesgue measure. From (14) and (12) we have

$$\int \frac{p_n^\varepsilon(x - \lambda a_n)}{p_n^\varepsilon(x)} \log \frac{p_n^\varepsilon(x - \lambda a_n)}{p_n^\varepsilon(x)} \cdot p_n^\varepsilon(x) \, m_n(dx) \leq$$
$$\leq \frac{1}{\delta} \left(e^{\frac{1}{\delta}\lambda} - 1\right) \int \exp\{\delta \, |l_\mu(a, x)|\} \, \mu(dx)$$

uniformly w.r.t. ε. Using this and the fact that for $f \in C_X$

$$\lim_{\varepsilon \to \infty} \int f(x) \, \bar{\mu}^\varepsilon(dx) = \int f(x) \, \mu(dx)$$

we can show that $\lambda a \in M_\mu$ and that (10) is satisfied.

§ 23. Quasi-invariant Measures

A measure μ on $(X, \mathfrak{B})$ is said to be *quasi-invariant* if the set M_μ of admissible shifts for it contains a linear manifold dense in X. If L is such a manifold, then from $\mu_a \ll \mu$ and $\mu_{-a} \ll \mu$ for $a \in L$ it follows that $\mu_a \sim \mu$ for all $a \in L$. Any vector a ($|a| = 1$) from L will be an admissible direction for the measure μ. For any finite-dimensional space $L_n \subset L$, the projection of μ on L_n will be equivalent to Lebesgue measure on L_n.

Let $\{e_k\}$ be some orthonormalized basis in X. Denote by L the linear hull of the vectors $\{e_k\}$ and by $\mathfrak{M}$ the set of all finite measures on $(X, \mathfrak{B})$ for which $L \in M_\mu$. The goal of this section is the description of the set $\mathfrak{M}$.

A measure $\mu \in \mathfrak{M}$ is called *extremal* if it cannot be represented as the sum of two mutually singular measures from $\mathfrak{M}$.

We note some properties of measures from $\mathfrak{M}$ and of extremal measures.

Lemma 1. *If ν and μ are measures from $\mathfrak{M}$ and ν' is the absolutely continuous component of ν w.r.t. μ, then $\nu' \in \mathfrak{M}$ and $\nu - \nu' \in \mathfrak{M}$ provided that $\nu \neq \nu'$.*

Proof. Assume $\nu'' = \nu - \nu'$ and $\nu'' \perp \mu$. If $a \in L$, then $\nu_a' + \nu_a'' \sim \sim \nu' + \nu''$ and $\nu_a'' \ll \nu' + \nu''$. Since $\nu_a'' \perp \mu_a \sim \mu$ and $\nu' \ll \mu$, we have $\nu_a'' \perp \nu'$ and $\nu_a'' \ll \nu$, i.e., $\nu'' \in \mathfrak{M}$. In exactly the same way, $\nu_a' \ll \nu' + \nu''$ and $\nu_a' \perp \nu''$ so that $\nu_a' \ll \nu'$ and $\nu' \in \mathfrak{M}$. □

Corollary 1. *If μ and ν are two extremal measures, then either $\mu \sim \nu$ or $\mu \perp \nu$.* Indeed, if this were not the case, then a representation $\mu = \mu' + \mu''$ would hold, where $\mu' \ll \nu$ and $\mu'' \perp \nu$ and this is impossible if μ is extremal.

Corollary 2. *If μ is extremal, then for all $a \in X$, either $\mu \sim \mu_a$ or $\mu \perp \mu_a$.* This follows from the fact that μ_a is also extremal.

Corollary 3. *If μ is extremal, $\nu \in \mathfrak{M}$ and $\nu \ll \mu$, then ν is also extremal and $\nu \sim \mu$.* Otherwise $\mu = \mu' + \mu''$, where $\mu' \ll \nu$ and $\mu'' \perp \nu$, μ', $\mu'' \in \mathfrak{M}$.

A $\mathfrak{B}$-measurable real-valued function $h(x)$ is called *invariant* for the measure $\mu \in \mathfrak{M}$, if for all $a \in L$ and for μ-almost all x $h(x + a) = h(x)$.

Lemma 2. *In order for a measure μ to be extremal, it is necessary and sufficient that any function $h(x)$ which is invariant for μ coincide μ-a.e. with a constant.*

Proof. Let $h(x)$ be invariant for μ and assume there exists an α such that

$$\mu(\{x: h(x) < \alpha\}) > 0 \qquad \text{and} \qquad \mu(\{x: h(x) \geq \alpha\}) > 0 .$$

Set $\varphi_1(x) = 0$ for $h(x) < \alpha$, $\varphi_2(x) = 0$ for $h(x) \geq \alpha$ with $\varphi_1(x) + \varphi_2(x) = 1$ and

$$\mu^i(A) = \int_A \varphi_i(x)\, \mu(dx) .$$

The functions $\varphi_i(x)$ are obviously also invariant for μ. Hence, for $a \in L$

$$\begin{aligned}\int f(x)\,\mu_a^i(dx) &= \int f(x+a)\,\mu_i(dx) = \\ &= \int f(x+a)\,\varphi_i(x)\,\mu(dx) = \\ &= \int f(x+a)\,\varphi_i(x+a)\,\mu(dx) = \\ &= \int f(x)\,\varphi_i(x)\,\varrho_\mu(a,x)\,\mu(dx) = \\ &= \int f(x)\,\varrho_\mu(a,x)\,\mu^i(dx)\,.\end{aligned}$$

Thus, $\mu_a^i \ll \mu^i$ so that $\mu^i \in \mathfrak{M}$. Moreover, $\mu^1 \perp \mu$. Since $\mu = \mu^1 + \mu^2$, μ will not be extremal.

Now let $\mu = \mu^1 + \mu^2$, where $\mu^i \in \mathfrak{M}$ and $\mu^1 \perp \mu^2$. Write $\varphi_i(x) = \frac{d\mu_i}{d\mu}(x)$. Almost everywhere (μ) we have for $a \in L$

$$\varphi_1(x)\,\varphi_2(x+a) = \varphi_1(x)\,\varphi_2(x) = \varphi_1(x+a)\,\varphi_2(x) = 0$$

(since $\mu^1 \perp \mu^2$, $\mu_a^1 \perp \mu^2$, $\mu_a^2 \perp \mu^1$) and $\varphi_1(x) + \varphi_2(x) = 1$. Hence,

$$\begin{aligned}0 &= [(\varphi_1(x) - \varphi_1(x+a)) + (\varphi_2(x) - \varphi_2(x+a))]^2 = \\ &= (\varphi_1(x) - \varphi_1(x+a))^2 + (\varphi_2(x) - \varphi_2(x+a))^2\,.\end{aligned}$$

We see that the $\varphi_i(x)$ are nonconstant invariant functions for the measure μ. □

Let L_n be a subspace spanned by $e_1, \ldots, e_n$, X^n the orthogonal complement of L_n and $\mathfrak{B}_n$, $\mathfrak{B}^n$ σ-algebras generated by cylinder sets with bases in L_n and X^n, resp. We will denote by $\mu(\cdot, \mathfrak{A}/x)$ the conditional measure for μ w.r.t. the σ-algebra $\mathfrak{A} \subset \mathfrak{B}$ on the measurable space $(X, \mathfrak{B}, \mu)$.

Lemma 3. *If $h(x)$ is an invariant bounded function for the measure μ, then for all n*

$$h(x) = \int h(y)\,\mu(dy, \mathfrak{B}^n/x) \qquad \text{a.e. (w.r.t. } \mu)\,.$$

Proof. It is enough to show that there exists a $\mathfrak{B}^n$-measurable function $\tilde{h}(x)$ which coincides μ-a.e. with $h(x)$. Let $\varphi_\varepsilon(y)$ be the density of a Gaussian measure in L_n with mean 0 and correlation operator εI_n, where I_n is the identity operator in L_n, w.r.t. m_n, Lebesgue measure in L_n.

Put

$$h_\varepsilon(x) = \int_{L_n} h(x+y)\,\varphi_\varepsilon(y)\,m_n(dy)\,.$$

The $\mathfrak{B}$-measurability of $h(x)$ implies that $h_\varepsilon(x) \to h(x)$ for $\varepsilon \to 0$ (in μ). Hence, we can choose a sequence ε_k such that $h_{\varepsilon_k}(x) \to h(x)$ μ-a.e. We note that for $a \in L_n$ the function

$$h_\varepsilon(x+a) = \int h(x+y)\,\varphi_\varepsilon(y-a)\,dy$$

is continuous in a. Moreover, $h_\varepsilon(x + a) = h_\varepsilon(x)$ for μ-almost all x. From this it follows that $h_\varepsilon(x + a)$ is constant as a function of a for μ-almost all x. Hence, for μ-almost all x $h_\varepsilon(x) = h_\varepsilon(P^n x)$, where P^n is projection onto X^n. The function

$$\tilde{h}(x) = \lim_{k\to\infty} h_{\varepsilon_k}(P^n x)$$

is defined for all x for which this limit exists and is the one we are seeking. □

Corollary 1. *Let* $\bar{\mathfrak{B}}^n$ *be the completion of* $\mathfrak{B}^n$ *w.r.t. the measure* μ *and* $\mathfrak{B}^\infty = \bigcap_n \bar{\mathfrak{B}}^n \cap \mathfrak{B}$. *Then any invariant function for* μ *is* $\mathfrak{B}^\infty$-*measurable.* This follows from the fact that by Lemma 3, an invariant function $h(x)$ will be $\bar{\mathfrak{B}}^n$-measurable since the bounded invariant function arc tan $h(x)$ is $\bar{\mathfrak{B}}^n$-measurable.

Corollary 2. *The function* $h(x)$ *is invariant for* μ *iff it is* $\mathfrak{B}^\infty$-*measurable.* The necessity follows from Corollary 1. The sufficiency is a consequence of the fact that for the $\mathfrak{B}^\infty$-measurable function $h(x)$ one can determine for each n a $\mathfrak{B}^n$-measurable function $h_n(x)$ such that $\mu(\{x: h(x) = h_n(x)\}) = 1$ and such that for each $\mathfrak{B}^n$-measurable function $h_n(x)$, $h_n(x + y) = h_n(x)$ for all $y \in L_n$.

Corollary 3. *A normalized measure* μ *is extremal iff* μ *assumes only the values* 0 *and* 1 *on* $\mathfrak{B}^\infty$. Indeed, only in this case will all $\mathfrak{B}^\infty$-measurable functions be equivalent to constants.

We can now give a complete description of all extremal normalized measures and of measures from $\mathfrak{M}$. We will ascribe $\mu \in \mathfrak{M}$ to the class $\mathfrak{R}$ if for each n one can find an $m > n$ such that for any $A \in \mathfrak{B}_n$ and $B \in \mathfrak{B}^m$

$$\mu(A \cap B) = \mu(A)\,\mu(B)\,.$$

is satisfied.

Theorem 1. *A measure is extremal iff it is equivalent to some measure from* $\mathfrak{R}$.

Proof. Necessity. Let μ be an extremal measure. For $n \leq m$ we denote by $\mu_n^m(\cdot/y)$ the conditional measure $\mu(\cdot, \mathfrak{B}_n/y)$ on the σ-algebra $\mathfrak{B}^m$ and by μ^m the restriction of μ to $\mathfrak{B}^m$. We will show that for all y (mod μ), $\mu_n^m(\cdot/y) \sim \mu^m$. It suffices to prove that for μ-almost all y $\mu_n^n(\cdot/y) \sim \mu^n$ since $\mu_n^m(\cdot/y)$ and μ^m are the restrictions of these measures to the same σ-algebra $\mathfrak{B}^m$. Let $\tilde{\mu}^n(A/y)$ be defined for $A \in \mathfrak{B}$, $A \in X^n$ and $y \in L_n$ by

$$\tilde{\mu}^n(A/y) = \mu_n^n(P_n^{-1} A/P_n^{-1} y)\,, \tag{1}$$

where P_n is projection onto L_n and P^n projection onto X^n; the left side of (1) makes sense since $\mu_n^n(\cdot/y)$ is $\mathfrak{B}_n$-measurable. Denote by $\tilde{\mu}^n$ the projection of μ onto X^n and by μ_n the projection of μ onto L_n. It is enough to show that for μ_n-almost all y

$$\tilde{\mu}^n(\cdot/y) \sim \tilde{\mu}^n .$$

From the results of § 20 (see (12) there), it follows that $\mu \sim \mu_n \times \tilde{\mu}^n$ (if we consider X as $L_n \times X^n$). Since for $A \subset L_n$, $A \in \mathfrak{B}$, $B \subset X^n$ and $B \in \mathfrak{B}$

$$\mu(A \times B) = \int_A \mu_n(dy)\, \tilde{\mu}^n(B/y) ,$$

and

$$\mu_n \times \tilde{\mu}^n(A \times B) = \int_A \mu_n(dy)\, \tilde{\mu}^n(B) ,$$

hold, to prove that $\tilde{\mu}^n(\cdot/y) \sim \mu^n$ for μ_n-almost all y we need only use Theorem 1, § 18.

The relation

$$\mu_n^m(\cdot/y) \sim \mu^m (\text{mod}\, \mu)$$

is established. Now let $\mu_n^\infty(\cdot/y)$ and μ^∞ be the restrictions of these measures to $\mathfrak{B}^\infty$. Then

$$\mu_n^\infty(\cdot/y) \sim \mu^\infty \qquad (\text{mod}\, \mu) .$$

On the basis of Corollary 3 of Lemma 3, the measure μ^∞ takes only the values 0 and 1 on $\mathfrak{B}^\infty$. But then the measure $\mu_n^\infty(\cdot/y)$, which is equivalent to it, coincides with μ^∞. That is,

$$\mu_n^\infty(\cdot/y) = \mu^\infty (\text{mod}\, \mu) .$$

If y is such that $\mu_n^n(\cdot/y) \sim \mu^n$, then for all $m \geq n$, $\mu_n^m(\cdot/y) \sim \mu^m$ and $\mu_n^\infty(\cdot/y) = \mu^\infty$. We denote the set of such y by C_n. For $y \in C_n$ we set

$$\varrho_n^m(x/y) = \frac{d\mu_n^m(\cdot/y)}{d\mu^m}(x) .$$

From Theorem 6, § 15 we have

$$\varrho_n^{m+1}(x/y) = \int \varrho_n^m(x'/y)\, \mu(dx', \mathfrak{B}^{m+1}/x) , \tag{2}$$

which holds for $y \in C_n$. Eq. (2) implies that for all $N > n$, the (finite) sequence of functions $g_k(x) = \varrho_n^{N-k}(x/y)$; $k = 1, \ldots, N - n$ is a martingale.

Using Remark 2 after Theorem 1, § 14, we can show that if $A_{(\beta_1, \beta_2)}$ ($\beta_2 > \beta_1 \geq 0$) is the set of those x for which the sequence $\varrho_n^m(x/y)$ crosses the interval (β_1, β_2) an infinite number of times for fixed n, x and y, then

$$\mu(A_{(\beta_1, \beta_2)}) = 0$$

since $A_{(\beta_1,\beta_2)} = \bigcap_k A^k_{(\beta_1,\beta_2)}$, where $A^k_{(\beta_1,\beta_2)}$ is the set of x's for which the sequence $\varrho^m_n(x/y)$ crosses the interval (β_1, β_2) for fixed n, x and y no fewer than k times and

$$\mu(A^k_{(\beta_1,\beta_2)}) \leq \frac{2}{\beta_2 + (k-1)(\beta_2 - \beta_1)} \int \varrho^m_n(x/y)\, dx = \frac{2}{\beta_2 + k(\beta_2 - \beta_1)} . \tag{3}$$

The inequality (3) follows from the fact that

$$A^k_{(\beta_1,\beta_2)} = \bigcup_{N=n+1}^{\infty} A^{k,N}_{(\beta_1,\beta_2)} ,$$

where $A^{k,N}_{(\beta_1,\beta_2)}$ is the set of x's for which the sequence $\varrho^n_n(x/y), \ldots, \varrho^N_n(x/y)$ crosses the interval (β_1, β_2) no fewer than k times. Remark 2 following Theorem 1, § 14 is applicable here since $\varrho^N_n(x/y), \varrho^{N-1}_n(x/y), \ldots, \varrho^n_n(x/y)$ is a martingale.

Using this fact again and Lemma 2, § 14, we convince ourselves that the sequence $\{\varrho^m_n(x/y); m = n, n+1, \ldots\}$ is bounded (mod μ). Hence, for $y \in C_n$

$$\lim_{m\to\infty} \varrho^m_n(x/y) \qquad (\text{mod } \mu)$$

exists. From (2) and Jensen's inequality, we find that for an arbitrary function ψ which is convex from below

$$\int \psi(\varrho^{m+1}_n(x/y))\, \mu(dx) \leq \int \psi(\varrho^m_n(x/y))\, \mu(dx) \leq \int \psi(\varrho^n_n(x/y))\, \mu(dx) .$$

Hence, the sequence $\varrho^m_n(x/y)$ is uniformly integrable w.r.t. m. Since for any $\mathfrak{B}^\infty$-measurable function $h(x)$

$$\lim_{m\to\infty} \int h(x)\, \varrho^m_n(x/y)\, \mu(dx) = \lim_{m\to\infty} \int h(x)\, \varrho^m_n\, (dx/y) =$$

$$= \int h(x)\, \mu^\infty_n(dx/y) = \int h(x)\, \mu(dx) ,$$

we have $\lim_{m\to\infty} \varrho^m_n(x/y) = 1$ (mod μ), $y \in C_0$. The uniform integrability of $\varrho^m_n(x/y)$ implies that

$$\lim_{m\to\infty} \int |\varrho^m_n(x/y) - 1|\, \tilde{\mu}^n(dx) = \lim_{m\to\infty} \int |\varrho^m_n(x/y) - 1|\, \mu(dx) = 0 . \tag{4}$$

On the basis of Theorem 1 § 18, we can assume that $\varrho^m_n(x/y)$ is $\mathfrak{B}_n \times \mathfrak{B}^m$-measurable. Integrating (4) w.r.t. the measure $\mu(dy)$ we find that

$$\lim_{m\to\infty} \iint |\varrho^m_n(x/y) - 1|\, \mu^n(dx)\, \mu(dy) = 0 . \tag{5}$$

We denote by $\hat{\mu}_n$ the product measure $\mu_n \times \tilde{\mu}^n$. Then from (5)

$$\lim_{m\to\infty} \iint |\varrho^m_n(x/x) - 1|\, \hat{\mu}_n(dx) = 0 . \tag{6}$$

By the definition of $\varrho_n^m(x/y)$ we have

$$\varrho_n^n(x/x) = \frac{d\mu}{d\hat{\mu}_n}(x) .$$

We introduce the conditional measures $\mu(\cdot, \mathfrak{B}^n/y)$. If $\tilde{\mu}_n$ and $\mu_n(\cdot, \mathfrak{B}^n/y)$ are the restrictions of the measures μ and $\mu(\cdot, \mathfrak{B}^n/y)$ to the σ-algebra $\mathfrak{B}_n$, then exactly as in the case of the measures $\mu_n^m(\cdot/y)$, one can establish that

$$\mu_n(\cdot, \mathfrak{B}^n/y) \sim \tilde{\mu}_n$$

for μ^n-almost all y and that

$$\frac{d\mu_n(\cdot, \mathfrak{B}^n/y)}{d\mu_n}(x) = \varrho_n^n(y/x) .$$

Let $l < n < m$ and assume $\hat{\nu}_{l,n}^m$ is the restriction of $\hat{\mu}_n$ to the σ-algebra $\mathfrak{B}^l \cap \mathfrak{B}_m$ and that ν_l^m is the restriction of μ to the same σ-algebra. Set

$$\pi_{l,n}^m(x) = \frac{d\nu_l^m}{d\hat{\nu}_{l,n}^m}(x) .$$

Then

$$\pi_{l,n}^m(x) = \int \varrho_n^n(x'/x')\, \hat{\mu}_n(dx', \mathfrak{B}^l \cap \mathfrak{B}_m/x) .$$

(We used Theorem 6, § 15 repeatedly). On the basis of Theorem 3, § 15 we can conclude that the limit

$$\lim_{m\to\infty} \pi_{l,n}^m(x) = \int \varrho_n^n(x'/x')\, \hat{\mu}_n(dx', \mathfrak{B}^l/x) = \frac{d\mu^l}{d\hat{\mu}_n^l}(x) , \tag{7}$$

exists $\hat{\mu}_n$-a.e., where $\hat{\mu}_n^l$ is the restriction of $\hat{\mu}_n$ to $\mathfrak{B}^l$. Assume $l < n < m < k$ and introduce the σ-algebra $\mathfrak{B}_{l,n}^{m,k}$ generated by sets of the form $A \cap B$, where $A \subset \mathfrak{B}^l \cap \mathfrak{B}_n$ and $B \subset \mathfrak{B}^m \cap \mathfrak{B}_k$. Put

$$f_{l,n}^{m,k}(x) = \int \frac{1}{\varrho_n^m(x'/x')}\, \mu(dx', \mathfrak{B}_{l,n}^{m,k}/x)\, \frac{\pi_{l,m}^k(x)}{\frac{d\mu^l}{d\hat{\mu}_m^l}(x)} . \tag{8}$$

From (6) and (7) it follows that for any fixed l and n

$$\lim_{m\to\infty} \lim_{k\to\infty} f_{l,n}^{m,k}(x) = 1 \qquad (\text{mod}\ \mu) . \tag{9}$$

We can write (8) as follows:

$$f_{l,n}^{m,k}(x) = \frac{d\hat{\mu}_m(\mathfrak{B}^l, \cdot)}{d\mu(\mathfrak{B}^l, \cdot)}(x) \cdot \frac{d\mu(\mathfrak{B}^l \cap \mathfrak{B}_k, \cdot)}{d\hat{\mu}_m(\mathfrak{B}^l \cap \mathfrak{B}_k, \cdot)}(x) \cdot \frac{d\hat{\mu}_n(\mathfrak{B}_{l,n}^{m,k}, \cdot)}{d\mu(\mathfrak{B}_{l,n}^{m,k}, \cdot)}(x) , \tag{10}$$

where $\hat{\mu}_m(\mathfrak{A}, \cdot)$ and $\mu(\mathfrak{A}, \cdot)$ are the restrictions of $\hat{\mu}_m$ and μ to the σ-algebra $\mathfrak{A} \subset \mathfrak{B}$. It is easy to see from (10) that

$$\int f_{l,n}^{m,k}(x)\, \mu(dx) = 1 .$$

If $n_1 < n_2 < \cdots < n_{2N+1}$ is an arbitrary sequence of numbers, then setting

$$\varphi_k(x) = f^{n_{2k},\, n_{2k+1}}_{n_{2k-2},\, n_{2k-1}}(x) \qquad (n_{2k-2} = 0 \text{ for } k = 1,\ \mathfrak{B}^0 = \mathfrak{B})$$

we have

$$\int \prod_{k=1}^{N} \varphi_k(x)\, \mu(dx) = \int \prod_{k=2}^{N} \varphi_k(x)\, \mu(\mathfrak{B}^{n_2}, dx) = \cdots = \int \varphi_N(x)\, \mu(\mathfrak{B}^{n_{2N-2}}, dx) = 1\,. \tag{11}$$

The function φ_k is μ-a.e. positive. From (9) it follows that one can choose the sequence n_k in such a way that the infinite product $\prod_{k=1}^{\infty} \varphi_k(x)$ converges μ-a.e. Let $g(x) = \prod_{k=1}^{\infty} \varphi_k(x)$ and introduce the measure μ^* by means of:

$$\mu^*(A) = \int_A g(x)\, \mu(dx)\,.$$

From (11)

$$\int g(x)\, \mu(dx) \leq 1\,,$$

so that μ^* is a finite measure. Since $g(x) > 0 \pmod{\mu}$, $\mu \sim \mu^*$. Let $\psi(x)$ be a $\mathfrak{B}_{n_{2k-1}}$-measurable bounded function and $\tilde{\psi}(x)$ a bounded $\mathfrak{B}^{n_{2k}}$-measurable function. We calculate

$$\int \psi_n(x)\, \tilde{\psi}(x)\, \mu^*(dx) = \int \psi(x)\, \tilde{\psi}(x) \prod_{j=1}^{\infty} \varphi_j(x)\, \mu(dx) =$$

$$= \int \psi(x) \prod_{j=1}^{k} \varphi_j(x)\, \tilde{\psi}(x) \prod_{j=k+1}^{\infty} \varphi_j(x)\, \mu(dx)\,.$$

The function $\tilde{\psi}(x) \prod_{j=k+1}^{\infty} \varphi_j(x)$ is measurable w.r.t. $\mathfrak{B}^{n_{2k}}$ and $z_{k+1}(x) = \psi(x) \prod_{j=1}^{k} \varphi_j(x)$ is $\mathfrak{B}_{n_{2k+1}}$-measurable. Hence, the function $\int \tilde{\psi}(x) \prod_{j=k+1}^{\infty} \varphi_j(x')\, \mu(dx', \mathfrak{B}_{n_{2k+1}}/x)$ is measurable w.r.t. $\mathfrak{B}^{n_{2k}} \cap \mathfrak{B}_{n_{2k+1}}$ and

$$\int \psi(x)\, \tilde{\psi}(x)\, \mu^*(dx) = \int \psi(x) \prod_{j=1}^{k} \varphi_j(x)\, z_{k+1}(x)\, \mu(\mathfrak{B}_{n_{2k+1}}/dx)\,. \tag{12}$$

Now set

$$\int \psi(x') \prod_{j=1}^{k-1} \varphi_j(x')\, \mu(dx', \mathfrak{B}^{n_{2k-2}}/x) = z^{k-1}(x)\,.$$

Since $\psi(x) \prod\limits_{j=1}^{k-1} \varphi_j(x)$ is $\mathfrak{B}_{n_{2k-1}}$-measurable, $z^{k-1}(x)$ will be $\mathfrak{B}^{n_{2k-2}} \cap$

$\cap \mathfrak{B}_{n_{2k-1}}$-measurable. Eq. (12) can now be written as

$$\int \psi(x)\, \tilde{\psi}(x)\, \mu^*(dx) =$$

$$= \int z^{k-1}(x)\, z_{k+1}(x) \cdot \frac{d\hat{\mu}_{n_{2k}}(\mathfrak{B}^{n_{2k-2}}, \cdot)}{d\mu(\mathfrak{B}^{n_{2k-2}}, \cdot)}(x) \times$$

$$\times \frac{d\mu\left(\mathfrak{B}^{n_{2k-2}} \cap \mathfrak{B}_{n_{2k+1}}\right)}{d\hat{\mu}_{n_{2k}}\left(\mathfrak{B}^{n_{2k-2}} \cap \mathfrak{B}_{n_{2k+1}}\right)}(x) \times$$

$$\times \frac{d\hat{\mu}_{2k-1}\left(\mathfrak{B}^{n_{2k},\, n_{2k+1}}_{n_{2k-2},\, n_{2k-1}}, \cdot\right)}{d\mu\left(\mathfrak{B}^{n_{2k},\, n_{2k+1}}_{n_{2k-2},\, n_{2k-1}}, \cdot\right)}(x)\, \mu(\mathfrak{B}_{n_{2k+1}}, dx) =$$

$$= \int z^{k-1}(x)\, z_{k+1}(x)\, d\hat{\mu}_{n_{2k-1}}(dx) = \int z^{k-1}(x)\, \mu(dx) \int z_{k+1}(x)\, \mu(dx) =$$

$$= \int \psi(x) \prod_{j=1}^{k-1} \varphi_j(x)\, \mu(dx) \cdot \int \tilde{\psi}(x) \prod_{j=k+1}^{\infty} \varphi_j(x)\, \mu(dx) .$$

Setting $\psi = 1$ and then $\tilde{\psi} = 1$ in the equality

$$\int \psi(x)\, \tilde{\psi}(x)\, \mu^*(dx) = \int \psi(x) \prod_{j=1}^{k-1} \varphi_j(x)\, \mu(dx) \cdot \int \tilde{\psi}(x) \prod_{j=k+1}^{\infty} \varphi_j(x)\, \mu(dx) \quad (13)$$

we get

$$\int \tilde{\psi}(x)\, \mu^*(dx) = \int \psi_i(x) \prod_{j=k+1}^{\infty} \varphi_j(x)\, \mu(dx) ,$$

and

$$\int \psi(x)\, \mu^*(dx) = \int \psi(x) \prod_{j=1}^{k-1} \varphi_j(x)\, \mu(dx)\, \mu^*(X) .$$

Hence,

$$\int \psi(x)\, \tilde{\psi}(x)\, \mu^*(dx) = \frac{1}{\mu^*(X)} \int \psi(x)\, \mu^*(dx) \cdot \int \tilde{\psi}(x)\, \mu^*(dx) . \quad (14)$$

If now $\tilde{\mu}(A) = \mu^*(A)/\mu^*(X)$, then $\mu \sim \tilde{\mu}$ and $\bar{\mu} \subset \mathfrak{R}$. Consequently, it follows from (14) that for any k and any pair of sets $A \in \mathfrak{B}_{n_{2k-1}}$ and $B \in \mathfrak{B}^{n_{2k}}$, we have

$$\bar{\mu}\,(A \cap B) = \int \chi_A(x)\, \chi_B(x)\, \bar{\mu}(dx) = \int \chi_A(x)\, \bar{\mu}(dx) \cdot \int \chi_B(x)\, \bar{\mu}(dx) =$$

$$= \bar{\mu}(A)\, \bar{\mu}(B) .$$

The necessity is proved.

Sufficiency. We will show that each measure $\mu \in \mathfrak{R}$ is extremal. Let $h(x)$ be a $\mathfrak{B}^\infty$-measurable function. Then, by the definition of the set $\mathfrak{R}$, for each bounded, $\mathfrak{B}_n$-measurable function $\psi(x)$

$$\int \psi(x)\, h(x)\, \mu(dx) = \int \psi(x)\, \mu(dx) \int h(x)\, \mu(dx) . \quad (15)$$

Relation (15) is satisfied on the set of functions $\psi(x)$ closed w.r.t. the bounded convergence of functions. Hence, (15) is correct for all bounded $\mathfrak{B}$-measurable functions. Setting $\psi = h$ in (15), we obtain

$$\int h^2(x)\, \mu(dx) = \left(\int h(x)\, \mu(dx)\right)^2 ,$$

whence

$$\int \left(h(x) - \int h(x)\, \mu(dx)\right)^2 \mu(dx) = 0 ,$$

i.e.,

$$h(x) = \int h(x)\, \mu(dx) \qquad (\mathrm{mod}\, \mu) .$$

Thus, each $\mathfrak{B}^\infty$-measurable function is equal to a constant (mod μ). It remains to use Lemma 2 and the corollaries to Lemma 3. □

We will now prove that a measure $\mu \in \mathfrak{M}$ can be obtained as a mixture of extremal measures (see § 18). In other words, one can determine a family of measures μ^y on $(X, \mathfrak{B})$, where $y \in (Y, \mathfrak{C})$, and a measure $\nu(\cdot)$ on $(Y, \mathfrak{C})$ such that

$$\mu = \int \mu^y\, \nu(dy) \tag{16}$$

for all $y \in Y$, $\mu^y \in \mathfrak{R}$. Formula (16) says that $\mu^y(B)$ is $\mathfrak{B}$-measurable for $B \in \mathfrak{B}$ and μ and μ^y are connected by (1) § 16. For the construction of the family of measures μ^y we identify $(Y, \mathfrak{C})$ with $(X, \mathfrak{B}^\infty)$ and take the restriction of μ to $\mathfrak{B}^\infty$ as ν. Moreover, we put

$$\mu^y(B) = \mu(B, \mathfrak{B}^\infty/y) ,$$

where $\mu(B, \mathfrak{B}^\infty/y)$ is the conditional measure for μ w.r.t. the σ-algebra $\mathfrak{B}^\infty$. Then, by (2) § 13 we have for the conditional measure

$$\mu(B) = \int \mu(B, \mathfrak{B}^\infty/y)\, \mu(dy) = \int \mu^y(B)\, \nu(dy) .$$

It remains to show that for ν-almost all y $\mu^y \in \mathfrak{R}$. We first prove that $\mu^y \in \mathfrak{M}$. Let $a \in L$. For any $\tilde{A} \in \mathfrak{B}^\infty$ and $f \in C_X$, we can write, using the invariance of the function $\chi_A(x)$

$$\iint_{\tilde{A}} f(x+a)\, \mu^y(dx)\, \nu(dy) = \int f(x+a)\, \chi_{\tilde{A}}(x)\, \mu(dx) =$$

$$= \int f(x)\, \chi_{\tilde{A}}(x)\, \varrho_\mu(a, x)\, \mu(dx) = \iint_{A} f(x)\, \varrho_\mu(a, x)\, \mu^y(dx)\, \nu(dy) .$$

Thus, for all $f \in C_X$ and ν-almost all y

$$\int f(x+a)\, \mu^y(dx) = \int f(x)\, \varrho_\mu(a, x)\, \mu^y(dx) . \tag{17}$$

Let C_0 be some countable set of functions $f \in C_X$ such that the closure of the linear hull of C_0 w.r.t. bounded convergence contains C_X. Since (17) holds for all $f \in C_0$ for ν-almost all y, it also holds for all $f \in C_X$

for ν-almost all y. Therefore,

$$\frac{d\mu_a^y}{d\mu^y}(x) = \varrho_\mu(a, x) \qquad (\text{mod } \nu) \tag{18}$$

and $\mu_a^y \sim \mu^y$ for all $a \in L$ and for ν-almost all y. Denote by m_1 Lebesgue measure on R^1 and let $m_1 \times \nu$ be a measure on the product $(R^1, \mathfrak{A}) \times \times (Y, \mathfrak{C})$, where $\mathfrak{A}$ is the σ-algebra of Borel sets on the line. Then, for almost all pairs $(\tau; y) \in R^1 \times Y$ (w.r.t. $m_1 \times \nu$) we have $\mu_{\tau a}^y \sim \mu^y$. Hence, for ν-almost all y the set $S_y \in \mathfrak{A}$ of all τ for which $\mu_{\tau a}^y \sim \mu^y$ (for fixed y) will have full Lebesgue measure (its complement has measure 0). Moreover, S_y is an additive group. This implies that $S_y = R^1$ for almost all y. Indeed, if this were not the case, then taking $\xi \notin S_y$ and setting $S_y + \xi = \{\tau : \tau - \xi \in S_y\}$ we would have: $S_y + \xi$ and S_y are disjoint and both have full Lebesgue measure.

We have this shown that for ν-almost all y each $a \in L$ is an admissible direction for the measure μ^y. This means that for ν-almost all y all e_k (generating L) are admissible directions for μ^y. We have hereby proved that $\mu^y \in \mathfrak{M}$ for almost all y.

We will now show that for all y (mod ν) $\mu^y \in \mathfrak{R}$. For fixed y let $\mu^y(\cdot, \mathfrak{B}^n/x)$ denote the conditional measure for μ^y w.r.t. the σ-algebra $\mathfrak{B}^n$. If $f \in C_X$ and $g_n(x)$ is $\mathfrak{B}^n$-measurable and bounded and $\tilde{A} \in \mathfrak{B}^\infty$, then

$$\begin{aligned}
&\int_{\tilde{A}} \int \int f(x)\, \mu^y(dx', \mathfrak{B}^n/x)\, g(x)\, \mu^y(dx)\, \nu(dy) = \\
&= \int_{\tilde{A}} \int f(x)\, g_n(x)\, \mu^y(dx)\, \nu(dy) = \int_{\tilde{A}} f(x)\, g_n(x)\, \mu(dx) = \\
&= \int_{\tilde{A}} g_n(x) \int f(x')\, \mu(x', \mathfrak{B}^n/x)\, \mu(dx) = \\
&= \int \int \chi_{\tilde{A}}(x)\, g_n(x) \int f(x')\, \mu(x', \mathfrak{B}^n/x)\, \mu^y(dx)\, \nu(dy) = \\
&= \int_{\tilde{A}} \int g_n(x) \int f(x')\, \mu(x', \mathfrak{B}^n/x)\, \mu^y(dx)\, \nu(dy)
\end{aligned} \tag{19}$$

since for all $\tilde{A} \in \mathfrak{B}^\infty$ $\mu^y(\tilde{A}) = \chi_{\tilde{A}}(y)$ (mod ν) (in fact, for all $\tilde{B} \in \mathfrak{B}^\infty$, $\mu(\tilde{A} \cap \tilde{B}) = \int_{\tilde{B}} \mu(\tilde{A}, \mathfrak{B}^\infty/y)\, \mu(dy) = \int_{\tilde{B}} \chi_{\tilde{A}}(y)\, \mu(dy)$). Because $\tilde{A}$ and $g_n(x)$ are arbitrary, by comparing the first and last terms in (19) we find that for ν-almost all y and $f \in C_X$

$$\int f(x')\, \mu^y(dx', \mathfrak{B}^n/x) = \int f(x')\, \mu(dx', \mathfrak{B}^n/x) \qquad (\text{mod } \mu^y)\,. \tag{20}$$

Since

$$\lim_{n \to \infty} \int f(x')\, \mu(dx', \mathfrak{B}^n/x) = \int f(x')\, \mu^x(dx') \qquad (\text{mod } \mu)\,,$$

for all $f \in C_0$, all m and all y (mod ν), we have

$$\int f(P_m x)\, \mu^y(dx) = \lim_{n \to \infty} \int f(P_m x')\, \mu^y(dx', \mathfrak{B}^n/x) \qquad (\text{mod } \mu^y)\,. \tag{21}$$

Hence, we can determine an $A' \in \mathfrak{B}^\infty$ with $\nu(A') = 1$ such that for $y \in A'$ (21) holds for all $f \in C_0$. Let $h(x)$ be a bounded function invariant for the measure μ^y, $y \in A'$. Since for each n $h(x)$ is measurable w.r.t. $\overline{\mathfrak{B}}^n(y)$, the completion of $\mathfrak{B}^n$ in the measure μ^y, we have for $m < n$

$$\int f(P_m x)\, h(x)\, \mu^y(dx) = \int h(x) \int f(P_m x')\, \mu^y(dx'; \mathfrak{B}^n/x)\, \mu^y(dx)\ .$$

Using (21), we get

$$\int f(P_m x)\, h(x)\, \mu^y(dx) = \int h(x) \int f(P_m x')\, \mu^y(dx')\, \mu^y(dx) = \\ = \int h(x)\, \mu^y(dx) \int f(P_m x)\, \mu^y(dx)$$

or

$$\int f(P_m x)\, [h(x) - \int h(x')\, \mu^y(dx')]\, \mu^y(dx) = 0\ .$$

Taking the properties of C_0 into account, we see that

$$h(x) = \int h(x')\, \mu^y(dx') \qquad (\text{mod}\ \mu^y)\ .$$

Hence, an arbitrary function h invariant for the measure μ^y is constant (mod μ^y). Therefore, for $y \in A'$ $\mu^y \in \mathfrak{R}$. We have proved

Theorem 2. *For any measure μ there exists a family of measures μ^y, $y \in (Y, \mathfrak{C})$, and a measure ν on $(Y, \mathfrak{C})$ such that $\mu^y \in \mathfrak{R}$ and* (16) *holds.*

Chapter 5. Some Questions of Analysis in Hilbert Space

§ 24. The Substitution Formula and Absolute Continuity

Let a measure μ be defined on $(X, \mathfrak{B})$ and let $T(x)$ be a $\mathfrak{B}$-measurable map from X into X. Then for any bounded $\mathfrak{B}$-measurable function $\varphi(x)$ we can define the integral

$$\int \varphi(T(x))\, \mu(dx)\,. \tag{1}$$

When X is a finite-dimensional Euclidean space, $T(x)$ is a differentiable invertible map and μ is Lebesgue measure, one can apply the substitution formula to the integral (1):

$$\int \varphi(T(x))\, \mu(dx) = \int \varphi(x)\, D(x)\, \mu(dx)\,, \tag{2}$$

where $D(x)$ is the modulus of the Jacobian of the inverse transformation of $T(x)$. The left side of (2) can be written as the integral $\int \varphi(x)\, \nu(dx)$, where the measure ν is defined by $\nu(A) = \mu(T^{-1}(A))$. Thus,

$$\int \varphi(x)\, \nu(dx) = \int \varphi(x)\, D(x)\, \mu(dx)\,, \tag{3}$$

i.e., the substitution formula says in this case that the measure ν, obtained from μ under the transformation T, is absolutely continuous w.r.t. μ and

$$\frac{d\nu}{d\mu}(x) = D(x)\,. \tag{4}$$

Clearly, (3) and (4) are equivalent. Hence, in order to be able to write the substitution formula in the form (2) for an arbitrary measure ν, we must establish that $\nu \ll \mu$ and find an expression for $\frac{d\nu}{d\mu}$.

It turns out that for a quasi-invariant measure μ one can find $\frac{d\nu}{d\mu}$ under a wide range of conditions. In order to find an expression for this derivative we will limit ourselves in this section to the finite-dimensional case. The justification of such an expression in infinite-dimensional spaces will be carried out in the following sections.

Thus, let $(X, \mathfrak{B})$ be a finite-dimensional space. If μ is quasi-invariant, it is equivalent to Lebesgue measure. Denote the density of μ

w.r.t. Lebesgue measure on X (we will denote the latter by m) by $p(x)$. If $T(x)$ is an invertible differentiable mapping of X into X and $S(x)$ is the inverse of T, then $S'(x)$ will stand for the derivative of S (which is a linear operator). Let $\det U$ be the determinant of the matrix of the linear operator U in an orthonormalized basis. Then

$$\int \varphi(T(x))\, \mu(dx) = \int \varphi(T(x))\, p(x)\, m(dx) =$$
$$= \int \varphi(y)\, p(S(y))\, |\det S'(y)|\, m(dy) =$$
$$= \int \varphi(y) \cdot \frac{p(S(y))}{p(y)}\, |\det S'(y)|\, \mu(dy)\,.$$

Hence, in this case

$$\frac{d\nu}{d\mu}(x) = \frac{p(S(x))}{p(x)} \cdot |\det S'(x)|\,. \tag{5}$$

We will transform the right side of (5) in such a way that it makes sense in an infinite-dimensional space. It is then natural to expect that (5) (possibly under additional restrictions) will also hold for infinite dimensional spaces.

We recall that a mapping $S(x)$ of X into X is differentiable (weakly) at a point x_0 if there exists a bounded linear operator $S'(x_0)$ (called the derivative of $S(x)$ at x_0) such that for all y and all $\lambda > 0$

$$|S(x_0 + \lambda y) - S(x_0) - \lambda S'(x_0)\, y| = o(\lambda)\,,$$

whereby $o(\lambda)$ is uniform w.r.t. y if $\|y\| < 1$ and $y \in L$, and L is an arbitrary finite-dimensional subspace of X. We now define the quantity $|\det U|$ for a class of linear operators U. If U is an orthogonal operator, we set $|\det U| = 1$. If U is symmetric with purely discrete spectrum and $\{\lambda_k\}$ are its eigenvalues, whereby each value in this sequence is repeated exactly as many times as its multiplicity, then

$$|\det U| = \prod_k |\lambda_k|\,,$$

provided this product converges, or diverges to zero (independently of the order of multiplication). Finally, for the product $U_1 U_2$ of two operators one of which is unitary and the other symmetric, we set

$$|\det U_1 U_2| = |\det U_1| \cdot |\det U_2|\,.$$

Using these definitions of the derivative and $|\det U|$ we can also give a meaning to the expression $|\det S'(y)|$ in an infinite-dimensional space.

We turn now to $\frac{p(S(y))}{p(y)}$. Note that for a measure μ in a finite-dimensional space X with positive density $p(x)$ w.r.t. Lebesgue measure,

$\varrho_\mu(x, a)$ is defined for all $a \in X$ when $\varrho_\mu(a, x) = \frac{p(x-a)}{p(x)}$. We can assume that $p(x)$ is a Borel function on X. Then $\varrho_\mu(a, x)$ will be a Borel function of its two arguments, so that

$$\varrho_\mu(x - S(x), x) = \frac{p(S(x))}{p(x)}$$

is also a Borel function. Hence, in this case $\frac{p(S(x))}{p(x)}$ can be expressed by means of the function $\varrho_\mu(a, x)$ which is well-defined in an infinite-dimensional space. In the sequel we will need a variant of $\varrho_\mu(a, x)$ which is $\mathfrak{B} \times \mathfrak{B}$-measurable in (a, x). The fact is that for each $a \in M_\mu$ the function $\varrho_\mu(a, x)$ is defined w.r.t. x only up to equivalence (mod μ). By modifying $\varrho_\mu(a, x)$ for each a on a set of μ-measure 0, we can alter it considerably due to the noncountability of the set M_μ. In the sequel we will understand by $\varrho_\mu(a, x)$ the following "principal" variant. Let $\{e_k\}$ be a fixed basis from M_μ, L_n the subspace generated by $e_1, \ldots, e_n$ and $p_n(x)$ the density of the projection of μ on L_n w.r.t. Lebesgue measure. We then set

$$\hat{\varrho}_\mu(a, x) = \lim_{n\to\infty} \frac{p_n(P_n x - P_n a)}{p_n(P_n x)}.$$

for all a and x for which this limit exists and $\hat{\varrho}_\mu(a, x) = 0$ otherwise. Since $p_n(P_n x - P_n a)/p_n(P_n x)$ is a $\mathfrak{B} \times \mathfrak{B}$-measurable function in (a, x), $\hat{\varrho}_\mu(a, x)$ will be also. Finally, we put

$$\varrho_\mu(a, x) = \begin{cases} \hat{\varrho}_\mu(a, x) \text{ if } \int \hat{\varrho}_\mu(a, x)\, \mu(dx) = 1; \\ \text{undefined, if } \int \hat{\varrho}_\mu(a, x)\, \mu(dx) < 1. \end{cases}$$

This will be the "principal" $\mathfrak{B} \times \mathfrak{B}$-measurable variant of $\varrho_\mu(a, x)$. Therefore, in the infinite-dimensional case, it is natural to expect that under certain assumptions, the measure ν, obtained from μ under a transformation $T(x)$ having a differentiable inverse transformation $S(x)$, will be absolutely continuous w.r.t. μ, with

$$\frac{d\nu}{d\mu}(x) = |\det S'(x)|\, \varrho_\mu\left(S(x) - x, x\right). \tag{6}$$

Eq. (6) makes sense if $|\det S'(x)|$ is defined for all x (mod μ) and the pair $\left(S(x) - x, x\right)$ falls within the domain of definition of $\varrho_\mu(a, x)$. Obviously,

$$\varrho_\mu\left(S(x) - x, x\right) = \lim_{n\to\infty} \frac{p_n(P_n S(x))}{p_n(x)} \qquad (\text{mod}\ \mu). \tag{7}$$

The validity of (6) will be investigated in the following sections. Sometimes it is necessary in the derivation of formulas of the type (6) to use a specially chosen basis. Then, in place of $\varrho_\mu(a, x)$ in (6) we have $\tilde{\varrho}_\mu(a, x)$ say, which is constructed exactly as $\varrho_\mu(a, x)$, but w.r.t. the other basis. It is clear that for all $a \in M_\mu$

$$\mu\left(\{x: \varrho_\mu(a, x) = \tilde{\varrho}_\mu(a, x)\}\right) = 1 .$$

However, it does not follow from this that $\tilde{\varrho}_\mu$ can be replaced by ϱ_μ in (6) since a depends on x. This can be done if

$$\mu\Big(\bigcup_{a \in M_\mu} \{x: \varrho_\mu(a, x) = \tilde{\varrho}_\mu(a, x)\}\Big) = 1 . \tag{8}$$

If (8) is satisfied for any other basis $\{\tilde{e}_k\}$ in M_μ, we will say that the function $\varrho_\mu(a, x)$ *depends regularly* on the basis.

We now give a simple condition ensuring that $\varrho_\mu(a, x)$ depends regularly on the basis.

Lemma 1. *Assume the following conditions on $\tilde{\varrho}_\mu(a, x)$ are satisfied for an arbitrary basis $\{\tilde{e}_k\} \in M_\mu$:*

1) *If $\tilde{\varrho}_\mu(a_k, x)$, $k = 1, \ldots, N$, is defined for a given x, then for all real $\lambda_1, \ldots, \lambda_N$, $\tilde{\varrho}_\mu(\sum \lambda_k a_k, x)$ is defined and continuous in $\lambda_1, \ldots, \lambda_N$.*

2) *One can choose an increasing sequence of subspaces $H_n \subset M_\mu$, $\bigcup H_n$ dense in X, and operators Q_n projecting (not necessarily orthogonally) onto H_n (H_n and Q_n are independent of the choice of the basis) and a set $B \in \mathfrak{B}$ (which can depend on the basis), such that for all a and x for which $\tilde{\varrho}_\mu(a, x)$ is defined and $x \notin B$ we have*

$$\lim_{n \to \infty} \tilde{\varrho}_\mu(Q_n x, x) = \tilde{\varrho}_\mu(a, x) .$$

Then $\varrho_\mu(a, x)$ depends regularly on the basis.

Proof. Choose a countable set $Z \subset M_\mu$ so that it is everywhere dense in each H_n. Let

$$B' = \bigcup_{a \in Z} \{x: \varrho_\mu(a, x) \neq \tilde{\varrho}_\mu(a, x)\} .$$

Then $\mu(B') = 0$. If $x \notin B'$, then Condition 1 says that $\tilde{\varrho}_\mu(a, x) = \varrho_\mu(a, x)$ on each of the subspaces H_n. Hence, if $x \notin B' \cup B$, where B is the set mentioned in Condition 2, then $\varrho_\mu(a, x) = \tilde{\varrho}_\mu(a, x)$ for all $a \in M_\mu$ by Condition 2. □

Remark. One can show that for a Gaussian measure $\varrho_\mu(a, x)$ depends regularly on the basis. For this measure, the conditions of Lemma 1 are satisfied if one takes for H_n the invariant subspaces of the correlation operator and Q_n are the orthogonal projection operators.

§ 25. Linear Transformations

We first consider the case $T(x) = U x$, where U is an invertible linear transformation for which $|\det U|$ is defined and does not vanish. This case is simpler since $|\det U^{-1}|$, coinciding with $|\det S'(x)|$, is constant.

Let U be an orthogonal operator. Since $U^* x - x$ (U^* is the conjugate operator) must belong to M_μ and M_μ belongs by Theorem 1, § 19 to $B^{1/2} X$, where B is a nuclear operator in X, the operator $U^* - I$ (I is the identity mapping) must be completely continuous. Then, one can find a sequence of finite-dimensional subspaces $\widetilde{L}_n$ such that $\widetilde{L}_n \subset \widetilde{L}_{n+1}$, $\widetilde{L}_n$ is invariant w.r.t. $U' - I$ and $\bigcup \widetilde{L}_n$ is dense in X. Since $(U^* - I) X \subset M_\mu$, $\widetilde{L}_n \subset M_\mu$. As in § 24, we denote by ν the image of μ under U. Let $\varphi(x)$ be a cylinder function of the form $\varphi(x) = g(\widetilde{P}_n x)$, where $g(x)$ is a bounded Borel function in $\widetilde{L}_n$ and $\widetilde{P}_n$ is projection onto $\widetilde{L}_n$. Then

$$\int \varphi(x)\, \nu(dx) = \int g(\widetilde{P}_n U x)\, \mu(dx) = \int g(U\, x)\, \mu_{\widetilde{L}_n}(dx) =$$
$$= \int g(x) \frac{\tilde{p}_n(U^* x)}{\tilde{p}_n(x)}\, \mu_{\widetilde{L}_n}(dx),$$

where $\tilde{p}_n(x)$ is the density of $\mu_{\widetilde{L}_n}$ w.r.t. Lebesgue measure in $\widetilde{L}_n$. Thus for $m < n$

$$\int g(\widetilde{P}_m x)\, \nu(dx) = \int g(\widetilde{P}_m x) \frac{\tilde{p}_n(\widetilde{P}_n U^* x)}{\tilde{p}_n(\widetilde{P}_n x)}\, \mu(dx)\,. \tag{1}$$

Denote by $\tilde{\varrho}_\mu(a, x)$ the function (defined for all x and a for which the limit exists)

$$\lim_{n\to\infty} \frac{\tilde{p}_n(\widetilde{P}_n(x - a))}{\tilde{p}_n(\widetilde{P}_n x)}\,, \tag{2}$$

provided that $a \in M_\mu$. Assume that for all x (mod μ) the pair $(x - U^* x, x)$ belongs to the domain of definition of $\tilde{\varrho}_\mu(u, x)$. Assume $g \geq 0$ and $\tilde{\varrho}_\mu(x - U^* x, x) > 0$ (mod μ). Letting $n \to \infty$ in (1) we get by Fatou's Lemma:

$$\int g(\widetilde{P}_m x)\, \nu(dx) \geq \int g(\widetilde{P}_m x)\, \tilde{\varrho}_\mu(x - U^* x, x)\, \mu(dx)\,. \tag{3}$$

Eq. (2) implies that for any nonnegative $\mathfrak{B}$-measurable function $\varphi(x)$,

$$\int \varphi(x)\, \nu(dx) > \int \varphi(x)\, \tilde{\varrho}_\mu(x - U^* x, x)\, \mu(dx)\,,$$

whence it follows that $\mu \ll \nu$. Since by (1)

$$\frac{d\mu_{\widetilde{L}_n}}{d\nu_{\widetilde{L}_n}}(\widetilde{P}_n x) = \frac{\tilde{p}_n(\widetilde{P}_n x)}{\tilde{p}_n(\widetilde{P}_n U^* x)}\,, \tag{4}$$

and by Theorem 3, § 13

$$\lim_{n\to\infty} \frac{d\mu_{\widetilde{L}_n}}{d\nu_{\widetilde{L}_n}}(\widetilde{P}_n x) = \frac{d\mu}{d\nu}(x) \qquad (\text{mod}\ \nu)\,,$$

we have

$$\frac{d\mu}{d\nu}(x) = \frac{1}{\tilde{\varrho}_\mu(x - U^* x, x)} \qquad (\mathrm{mod}\ \mu) \tag{5}$$

(the right side of (4) converges (mod μ) to the right side of (5)). Hence, if $\frac{d\nu}{d\mu}$ denotes the density of the absolutely continuous component of ν w.r.t. μ, then

$$\frac{d\nu}{d\mu}(x) = \tilde{\varrho}_\mu(x - U^* x, x)\,. \tag{6}$$

The considerations above also yield a simple condition for $\nu \ll \mu$. Assume that for μ-all x the pair $(x - U x, x)$ belongs to the domain of definition of $\tilde{\varrho}_\mu(a, x)$. Then, denoting by ν^* the image of the measure μ under U^*, we get $\mu \ll \nu^*$ provided that $\tilde{\varrho}_\mu(x - U x, x) > 0$ (mod μ). Since μ goes into ν under U and ν^* into μ (U^* is the inverse of U), we get $\nu \ll \mu$. We have therefore proved

Theorem 1. *If U is an orthogonal transformation for which $x - U^* x \in M_\mu$ and $x - U x \in M_\mu$, the function $\tilde{\varrho}_\mu(a, x)$ is defined by (2), for all x (mod μ) the pairs $(x - U x, x)$ and $(x - U^* x, x)$ belong to the domain of definition of $\tilde{\varrho}_\mu(a, x)$ and*

$$\tilde{\varrho}_\mu(x - U^* x, x) > 0\,, \qquad \tilde{\varrho}_\mu(x - U x, x) > 0 \qquad (\mathrm{mod}\ \mu)\,,$$

then $\nu \sim \mu$ and (6) holds.

If $\varrho_\mu(a, x)$ depends regularly on the basis, then one can write $\varrho_\mu(a, x)$ in (6).

Now assume that U is a continuous symmetric operator. Let $\tilde{L}_n$ be its invariant subspaces. Without altering the arguments carried out in the case of an orthogonal operator, we find, in analogy to (1)

$$\int g(\tilde{P}_m x)\,\nu(dx) = \int g(\tilde{P}_m x)\,\frac{\tilde{p}_n(\tilde{P}_n U x)}{\tilde{p}_n(\tilde{P}_n x)}\,\delta_n\,\mu(dx)\,, \tag{7}$$

where $\delta_n = \prod_{k=1}^{n} |\lambda_k|$ and λ_k are the eigenvalues of U corresponding to the eigenvectors $\tilde{e}_k$, $k = 1, \ldots, n$, generating the subspace $\tilde{L}_n$. From this formula, we obtain in exactly the same way as above

Theorem 2. *If the operator U is symmetric and invertible, $U - I$ is completely continuous and $|\det U|$ is defined, and if for all x (mod μ), the pairs $(x - U x, x)$ and $(x - U^{-1} x, x)$ belong to the domain of definition of $\tilde{\varrho}_\mu(a, x)$ and*

$$\tilde{\varrho}_\mu(x - U x, x) > 0\,, \qquad \tilde{\varrho}_\mu(x - U^{-1} x, x) > 0 \qquad (\mathrm{mod}\ \mu)$$

then $\nu \sim \mu$ and

$$\frac{d\nu}{d\mu}(x) = |\det U|\,\tilde{\varrho}_\mu(x - U^{-1} x, x)\,. \tag{8}$$

Moreover, if $\varrho_\mu(a, x)$ depends regularly on the basis, then we can replace $\tilde{\varrho}_\mu$ by ϱ_μ in the formulation of the theorem and in (8).

Remark 1. Investigating the proof of Theorem 1 (presented before its statement), we note that the condition $\tilde{\varrho}_\mu(x - U x, x) > 0 \pmod{\mu}$ is only necessary to establish that $\nu \ll \mu$ ($\mu \ll \nu$ follows from the fact that $\tilde{\varrho}_\mu(x - U^{-1} x, x) > 0 \pmod{\mu}$). If the measure μ is such that $\mu \ll \nu$ implies $\mu \sim \nu$, then the condition $\tilde{\varrho}_\mu(x - U x, x) > 0 \pmod{\mu}$ becomes superfluous. This will happen, for example, if μ is an extremal measure (in particular, if it is Gaussian).

We can now formulate a general theorem from which we can extract a condition under which $\tilde{\varrho}_\mu$ can be replaced by ϱ_μ in Theorems 1 and 2 without requiring that ϱ_μ depend regularly on the basis. To this end, it is necessary to impose some additional restrictions on ϱ_μ. However, the conditions in the new formulations have the advantage that they can be more easily verified; in particular, it is not necessary to consider the $\tilde{\varrho}_\mu$, defined in other bases.

Theorem 3. *Let $\varrho_\mu(a, x)$ be defined as in § 24 and let L_n be the corresponding sequence of finite-dimensional subspaces. Denote by Q_n some sequence of operators projecting (not necessarily orthogonally) onto L_n. Assume the following conditions are fulfilled:*

a) *for each finite-dimensional subspace $L \subset M_\mu$, the set of x's for which $\varrho_\mu(a, x)$ is defined for $a \in L$ and is continuous in a has measure 1;*

b) *one can find a set B such that $\mu(B) = 0$ and when $x \notin B$*

$$\lim_{n\to\infty} \varrho_\mu(Q_n a, x) = \varrho_\mu(a, x) ,$$

for all a for which $\varrho_\mu(a, x)$ is defined; moreover, for each finite-dimensional subspace $L \subset M_\mu$ and all x $\pmod{\mu}$, the convergence is uniform for $a \in L \cap \{a: |a| \le c\}$ for all $c > 0$;

c) *the operator $U - I$ is completely continuous, U is invertible, $|\det U|$ is defined and* $\lim_{n\to\infty} |\det (I + Q_n (U - I))| = |\det U|$;

d) *for all x* $\pmod{\mu}$, *$\varrho_\mu(x - U^{-1} x, x)$ and $\varrho_\mu(x - U x, x)$ are defined and positive.*

Then $\nu \sim \mu$ and

$$\frac{d\nu}{d\mu}(x) = |\det U| \, \varrho_\mu(x - U^{-1} x, x) . \tag{9}$$

Proof. We first assume that U is such that the range of $U - I$ and $U^* - I$ coincides with the finite-dimensional subspace L. Then, writing

$$U = U_1 U_2 , \qquad U_2 = (U^* U)^{1/2} \qquad \text{and} \qquad U_1 = U U_2^{-1} ,$$

where U_1 is orthogonal and U_2 is symmetric, we see that $U_1 - I$, $U_2 - I$, $U_1^* - I$ and $U_2^{-1} - I$ have the same range L. We show that if U_1 is an orthogonal operator for which $U_1 - I$ has range L, then we have for U_1 the formula

$$\int f(U_1 x)\, \mu(dx) = \int f(x)\, \varrho_\mu(x - U_1^{-1} x, x)\, \mu(dx) \tag{10}$$

for any $\mathfrak{B}$-measurable bounded function f. If L coincides with one of the subspaces L_n, Eq. (10) follows from Theorem 1 ($\tilde{\varrho}_\mu = \varrho_\mu$ and $\varrho_\mu(a, x)$ is positive on finite-dimensional subspaces). Choose a sequence of orthogonal operators V_n for which $V_n - I$ has range $Q_n[L]$ and

$$\lim_{n\to\infty} \|V_n - U_1\| = 0 .$$

Then, for all n

$$\int f(V_n x)\, \mu(dx) = \int f(x)\, \varrho_\mu(x - V_n^{-1} x, x)\, \mu(dx) . \tag{11}$$

Obviously,

$$\|(x - V_n^{-1} x) - (x - U_1^{-1} x)\| \leq \|V_n^{-1} - U_1^{-1}\| \cdot \|x\| \to 0$$

for all x. Moreover, for sufficiently large n, the dimension of $Q_n[L]$ is the same as that of L and we define the operator $\tilde{Q}_n$ transferring L into $Q_n[L]$ as follows: for $x \in L$ $\tilde{Q}_n x = Q_n x \in Q_n[L]$ and $\lim\limits_{n\to\infty} \sup\limits_{\|x\|<1, x\in L} \|Q_n x - x\| = 0$, so that for sufficiently large n this operator is invertible. Hence,

$$\lim_{n\to\infty} \tilde{Q}_n^{-1} (x - V_n^{-1} x) = x - U_1^{-1} x .$$

Using Conditions a) and b) of the theorem we see that for all x (mod μ)

$$\lim_{n\to\infty} \varrho\, (x - V_n^{-1} x, x) = \lim_{n\to\infty} \varrho_\mu\, (Q_n\, [\tilde{Q}_n^{-1}\, (x - V_n^{-1} x)], x) =$$

$$= \lim_{n\to\infty} \varrho\, (\tilde{Q}_n^{-1}\, [x - V_n^{-1} x], x) = \varrho(x - U_1^{-1} x, x) .$$

Let the function f in (11) be nonnegative and continuous. Letting $n \to \infty$, we get by Fatou's lemma

$$\int f(U_1 x)\, \mu(dx) \geq \int f(x)\, \varrho_\mu(x - U_1^{-1} x, x)\, \mu(dx) . \tag{12}$$

Arguing in exactly the same way we find that

$$\int f(U_1^{-1} x)\, \mu(dx) \geq \int f(x)\, \varrho_\mu(x - U_1 x, x)\, \mu(dx) . \tag{13}$$

Inequalities (12) and (13) hold for all nonnegative $\mathfrak{B}$-measurable f. Since

$$f(x)\, \varrho_\mu\, (x - U_1^{-1} x, x) = \Phi(U_1^{-1} x) ,$$

where

$$\Phi(x) = f(U_1 x)\, \varrho_\mu(U_1 x - x, U_1 x) ,$$

applying (13) to Φ, we have

$$\int f(x)\, \varrho_\mu(x - U_1^{-1} x, x)\, \mu(dx) \geq$$
$$\geq \int f(U_1 x)\, \varrho_\mu(U_1 x - x, U_1 x)\, \varrho_\mu(x - U_1 x, x)\, \mu(dx)\,. \tag{14}$$

But for all $a \in M_\mu$ and x (mod μ)

$$\varrho(a, x) \cdot \varrho(-a, x - a) = 1$$

holds. Substituting $x - U_1 x$ for a and using the fact that $x - U_1 x \in L$ along with Condition a) of the theorem, we can show that

$$\varrho_\mu(U_1 x - x, U_1 x)\, \varrho_\mu(x - U_1 x, x) = 1 \qquad (\mathrm{mod}\ \mu)\,.$$

Inserting this equality on the right of (14) and comparing with (12), we arrive at (10) for nonnegative functions f. The extension to f of arbitrary sign is obvious.

Now let U_2 be a symmetric operator whose range coincides with L. Arguing exactly as before we get the formula

$$\int f(U_2 x)\, \mu(dx) = \int f(x)\, \varrho_\mu(x - U_2^{-1} x, x) |\det U_2|\, \mu(dx)\,. \tag{15}$$

From (10) and (15) it follows that for any linear operators U_1 and U_2 for which $U_1 - I$ and $U_2 - I$ have finite-dimensional ranges with U_1 orthogonal and U_2 symmetric, we can write

$$\int f(U_1 U_2 x)\, \mu(dx) = \int f(U_1 x)\, \varrho_\mu(x - U_2^{-1} x, x)\, |\det U_2|\, \mu(dx) =$$
$$= \int f(x)\, \varrho_\mu(U_1^{-1} x - U_2^{-1} U_1^{-1} x, U_1^{-1} x)\, |\det U_2|\, \varrho_\mu(x - U_1^{-1} x, x)\, \mu(dx)\,.$$

Using the equality

$$\varrho_\mu(a + b, x) = \varrho_\mu(a, x)\, \varrho_\mu(b, x - a) \qquad (\mathrm{mod}\ \mu)\,,$$

the fact that $U_1^{-1} x - U_2^{-1} U_1^{-1} x$ and $x - U_1^{-1} x$ lie in finite-dimensional subspaces and Property a), we obtain

$$\varrho_\mu(U_1^{-1} x - U_2^{-1} U_1^{-1} x, U_1^{-1} x)\, \varrho_\mu(x - U_1^{-1} x, x) =$$
$$= \varrho_\mu(x - U_2^{-1} U_1^{-1} x, x) \qquad (\mathrm{mod}\ \mu)\,.$$

Hence,

$$\int f(U_1 U_2 x)\, \mu(dx) = \int f(x)\, \varrho_\mu\big(x - (U_1 U_2)^{-1} x, x\big)\, |\det U_1 U_2|\, \mu(dx)\,.$$

We have thus established (9) for the case in which U is such that the range of the operator $U - I$ is finite-dimensional.

Now let U be an arbitrary operator satisfying the conditions of the theorem. Set $U_n = I + Q_n (U - I)$. The operator U_n is such that the range of $U_n - I$ lies in L_n. Thus, according to what has been proved

$$\int f(U_n^{-1} x)\, \mu(dx) = \int f(x)\, \varrho_\mu(Q_n x - Q_n U x, x) \det |U_n^{-1}|\, \mu(dx)\,. \tag{16}$$

Since $U - I$ is a completaly continuous operator, Q_n is totally bounded, and for all x $\lim_{n\to\infty} Q_n x = x$, one has $\|U - U_n\| = \|(I - Q_n)(U - I)\| \to 0$ for $n \to \infty$. Hence $U_n^{-1} x \to U^{-1} x$. Choosing f continuous und nonnegative we find using b), c) and d) that

$$\int f(U^{-1} x)\, \mu(dx) \geq \int f(x)\, \varrho_\mu(x - U x, x)\, |\det U|^{-1}\, \mu(dx)\,. \qquad (17)$$

In an analogous way we can show that for nonnegative continuous functions f

$$\int f(U x)\, \mu(dx) \geq \int f(x)\, \varrho_\mu(x - U^{-1} x, x)\, |\det U|\, \mu(dx)\,. \qquad (18)$$

Using the same procedure with whose help (10) was obtained from (12) and (14), we calculate from (17) and (18) the equality

$$\int f(U x)\, \mu(dx) = \int f(x)\, \varrho_\mu(x - U^{-1} x, x)\, |\det U|\, \mu(dx)\,, \qquad (19)$$

which is valid for all nonnegative continuous f. This equality implies that $\nu \ll \mu$ and the validity of (9). Since $\frac{d\nu}{d\mu} > 0 \pmod{\mu}$, $\nu \sim \mu$. □

As an example of the application of Theorem 3 we consider the case in which the measure μ is Gaussian with mean 0 and correlation operator B. Denote by $\{e_k\}$ the sequence of eigenvectors and by $\{\lambda_k\}$ the corresponding eigenvalues of the operator B. Let the L_n appearing in the definition of $\varrho_\mu(a, x)$ be subspaces generated by $e_1, \ldots, e_n$. Then

$$M_\mu = \left\{a : \sum \frac{(a, e_k)^2}{\lambda_k} < \infty\right\},$$

and

$$\varrho_\mu(a, x) = \exp\left\{h(x, a) - \frac{1}{2}\xi(a)\right\},$$

where

$$h(x, a) = \sum_{k=1}^{\infty} \frac{(x, e_k)\,(a, e_k)}{\lambda_k} \qquad \text{and} \qquad \xi(a) = \sum_{k=1}^{\infty} \frac{(a, e_k)^2}{\lambda_k}.$$

The function $h(x, a)$ is defined for all (x, a) for which the corresponding series converges and the function $\xi(a)$ for all $a \in M_\mu$. It is clear that if for given x the function $h(x, a)$ is defined for $a_1, a_2, \ldots, a_n$, then it will also be defined for $\sum_{k=1}^{n} \beta_k a_k$ for any real $\beta_1, \ldots, \beta_n$ and

$$h\left(x, \sum_1^n \beta_k a_k\right) = \sum_1^n \beta_k h(x, a_k)\,.$$

This implies that $\varrho_\mu(a, x)$ satisfies Condition a). Choose as Q_n the orthogonal projection operators onto L_n. Then

$$\varrho_\mu(Q_n a, x) = \exp\left\{\sum_{k=1}^{n} \frac{(x, e_k)\,(a, e_k)}{\lambda_k} - \frac{1}{2}\sum_1^n \frac{(a, e_k)^2}{\lambda_k}\right\}.$$

If a belongs for given x to the domain of definition of ϱ_μ, then by the definition of the latter

$$\varrho_\mu(a, x) = \lim_{n\to\infty} \varrho_\mu(Q_n a, x) .$$

If for given x $\varrho_\mu(a, x)$ is defined for all $a \in L$, where L is some finite-dimensional subspace in M_μ, then for $a \in L \cap \{a: |a| < \alpha\}$

$$\lim_{n\to\infty} h(x, P_n a) = h(x, a) \quad \text{and} \quad \lim_{n\to\infty} \xi(P_n a) = \xi(a)$$

uniformly w.r.t. a since, choosing in L an orthonormalized basis $a_1, a_2, \ldots, a_m$, we will have

$$h(x, P_n a) = \sum_{k=1}^{m} (a, a_k)\, h(x, P_n a_k) ,$$

$$\xi(P_n a) = \sum_{k=1}^{n} \frac{\left(\sum_{j=1}^{m} (a, a_j)(a_j, e_k)\right)^2}{\lambda_k} =$$

$$= \sum_{i,j=1}^{m} (a, a_j)(a, a_i) \sum_{k=1}^{n} \frac{(a_i, e_k)(a_j, e_k)}{\lambda_k} ,$$

and the limit

$$\lim_{n\to\infty} \sum_{k=1}^{n} \frac{(a_i, e_k)(a_j, e_k)}{\lambda_k} = \lim_{n\to\infty} \frac{1}{2} \sum_{k=1}^{n} \frac{(a_i + a_j, e_k)^2 - (a_i, e_k)^2 - (a_j, e_k)^2}{\lambda_k}$$

exists. Therefore, Condition b) of Theorem 3 also holds.

Finally, we consider the consequences of Condition d) of Theorem 3. From the definition of $h(x, a)$ it follows that this function possesses the following two properties: 1) for each a, its domain of definition (as function of x) is a linear manifold and it is an additive homogeneous function of x on this manifold; 2) for all $a \in M_\mu$ the function $h(x, a)$ is defined for $x \in M_\mu$. These two properties imply that if $\varrho_\mu(x, a)$ is defined, then for all $b \in M_\mu$, $\varrho_\mu(a, x + b)$ is also defined; if $\varrho_\mu(a, x)$ and $\varrho_\mu(b, x)$ are defined, then $\varrho_\mu(a + b, x) = \varrho_\mu(a, x)\, \varrho_\mu(b, x - a)$ is also defined (the latter holds for any measure μ if $\varrho_\mu(a, x)$ and $\varrho_\mu(b, x - a)$ are defined; this is easy to verify for the Gaussian measure).

Now assume that the linear operator U is such that $U x - x \in M_\mu$ for all x. Then also $U^{-1} x - x \in M_\mu$ for all x. If the pair $(x - U^{-1} x, x)$ belongs to the domain of definition of the function $\varrho_\mu(a, x)$, then this means that for the given x the series

$$\sum_{k=1}^{\infty} \frac{(x - U^{-1} x, e_k)^2}{\lambda_k} \quad \text{and} \quad \sum_{k=1}^{\infty} \frac{(x - U^{-1} x, e_k)(x, e_k)}{\lambda_k}$$

converge. But the convergence of the first series follows from the fact that $x - U^{-1} x \in M_\mu$. Hence, $\varrho_\mu(x - U^{-1} x, x)$ is defined if for the given

x the series

$$\sum_{k=1}^{\infty} \frac{(x - U^{-1} x, e_k)(x, e_k)}{\lambda_k}. \tag{20}$$

converges. If $\varrho_\mu(x - U x, x)$ is defined, then it is necessarily positive. The function $\varrho_\mu(x - U x, x)$ is defined at x if

$$\sum_{k=1}^{\infty} \frac{(x - U x, e_k)(x, e_k)}{\lambda_k} \tag{21}$$

converges. We note that nothing is changed in the proof of Theorem 3 if L_n is an arbitrary nonincreasing sequence of finite-dimensional subspaces (i.e., the dimension of L_n does not necessarily equal n). That is, in the definition of $h(x, a)$ we can require that some subsequence of partial sums of the defining series converge (the same subsequence for all x). Hence, we can also consider convergence w.r.t. a subsequence in the series (20) and (21). Such a subsequence can be found (the same one for all x) if (20) and (21) converge w.r.t. the measure μ.

We thus have

Theorem 4. *Let μ be Gaussian with mean* 0 *and correlation operator B whose eigenvectors are $\{e_k\}$ and eigenvalues λ_k. Let the operator U satisfy:*

1. *U is invertible, $U - I$ is completely continuous,* det U *is defined and*

$$|\det U| = \lim_{n\to\infty} |\det (I - P_n (U - I))|,$$
$$|\det U^{-1}| = \lim_{n\to\infty} |\det (I - P_n (U^{-1} - I))|,$$

where P_n is projection onto the subspace L_n generated by the vectors $e_1, \ldots, e_n$;

2. *the series* (20) *and* (21) *converge in μ and for all x* (mod μ)

$$\sum_{k=1}^{\infty} \frac{(x - U^{-1} x, e_k)}{\lambda_k} < \infty \quad \textit{and} \quad \sum_{k=1}^{\infty} \frac{(x - U x, e_k)^2}{\lambda_k} < \infty .$$

Then, for the measure $\nu(A) = \mu(U^{-1}(A))$ we have: $\mu \sim \nu$ and

$$\frac{d\nu}{d\mu}(x) = |\det U| \cdot \exp\left\{\sum_{k=1}^{\infty} \frac{(x - U^{-1} x, e_k)(x, e_k)}{\lambda_k} - \frac{1}{2} \sum_{k=1}^{\infty} \frac{(x - U^{-1} x, e_k)^2}{\lambda_k}\right\}. \tag{22}$$

Remark. The measure ν considered in Theorem 4 is also Gaussian with mean 0 and correlation operator B_1 defined by

$$(B_1 z, z) = \int (z, z)^2 \nu(dx) = \int (z, U x)^2 \nu(dx) = \int (U^* z, x)^2 \mu(dx) =$$
$$= (B U^* z, U^* z).$$

This means that $B_1 = U B U^*$. Since μ and ν are Gaussian, necessary and sufficient conditions that $\mu \sim \nu$ and an expression for $\frac{d\nu}{d\mu}$ can be extracted from the results of § 17. These conditions will be more general (for example, it is not necessary to require that det U exist; the operator $(U - I)$ must be a Hilbert-Schmidt operator). However, the results used in the proof of Theorem 4 are essential for the derivation of formulas for nonlinear transformations in the following section.

§ 26. Absolute Continuity of Measures under Nonlinear Transformations

In this section we will use the results of Theorem 3 § 25 to derive Formula (6), § 24. The main result is contained in

Theorem 1. *Let the measure μ be such that there exists a linear manifold M_μ of admissible shifts of the measure μ and a function $\varrho_\mu(a, x)$, defined as in § 24 which satifies following conditions:*

1. *for each finite-dimensional subspace $L \subset M_\mu$ the measure of the set of those x for which $\varrho_\mu(a, x)$ is defined and continuous in a for $a \in L$ is equal to 1;*

2. *there exists a sequence of subspaces L_n and a set $B \in \mathfrak{B}$, $\mu(B) = 0$, such that for $x \notin B$ and all a for which $\varrho_\mu(a, x)$ is defined*

$$\lim_{n\to\infty} \varrho_\mu(P_n a, x) = \varrho_\mu(a, x)\,,$$

where P_n is projection onto L_n; moreover, for any finite-dimensional subspace L and $\alpha > 0$, the convergence takes place uniformly for $a \in L \cap \cap \{a\colon |a| < \alpha\}$ and all x (mod μ).

In addition, let $U(x)$ be a continuous, invertible, continuously differentiable transformation of X into X for which the following conditions hold:

3. *for all x,* det $U'(x)$ *exists* ($|\det U'(x)| \neq 0$), *the transformations $V_n x = x + P_n\,(U^{-1}(x) - x)$, and $U_n(x) = x + P_n\,(U(x) - x)$ are invertible and*

$$\lim_{n\to\infty} |\det U_n x| = |\det U'(x)|\,, \quad \lim_{n\to\infty} |\det V'_n(x)| = \frac{1}{|\det U'(x)|}\,;$$

4. *for all x* (mod μ), *$x - U(x) \in M_\mu$ and $x - U^{-1}(x) \in M_\mu$ and the quantities*

$$\varrho_\mu(x - U^{-1}(x), x) \quad \textit{and} \quad \varrho_\mu(x - U(x), x)$$

are defined and positive (mod μ).

Then, the measure ν, defined by $\nu(A) = \mu(U^{-1}(A))$, is equivalent to μ and

$$\frac{d\nu}{d\mu}(x) = |\det U'(U^{-1}x)|\, \varrho_\mu(x - U^{-1}(x), x) . \tag{1}$$

Proof. We first establish (1) for the case in which the operator $U(x)$ is such that for all x $U(x) - x \in L$, where L is some finite-dimensional subspace belonging to M_μ. We call a transformation $U(x)$ *polygonal* if the space X can be decomposed into a finite number of closed sets $D_1, \ldots, D_m$ having no common interior points in such a way that for any $k = 1, \ldots, m$ we can find a vector d_k and a linear operator V_k for which $U(x) = d_k + V_k x$, $x \in D_k$; $k = 1, \ldots, m$. It is easy to see that when a polygonal operator $U(x)$ is invertible, then $U^{-1}(x)$ is also polygonal, $U(x)$ will be differentiable at all interior points of D_k and

$$U'(x) = V_k , \qquad x \in D_k .$$

Let $U(x)$ be a polygonal operator for which $U(x) - x \in L$. Then $d_k \in L$ and $V_k x - x \in L$ for all x. Hence, for any $\mathfrak{B}$-measurable bounded function f and arbitrary k we have by Theorem 3, § 25

$$\int f(d_k + V_k x)\, \mu(dx) = \int f(d_k + x)\, \varrho_\mu(x - V_k^{-1} x, x)\, |\det V_k|\, \mu(dx) .$$

Using the fact that $d_k \in M_\mu$ we can rewrite this formula as

$$\int f(d_k + V_k x)\, \mu(dx) = \int f(x)\, \varrho_\mu(x' - V_k^{-1} x', x')\, |\det V_k|\, \varrho_\mu(d_k, x)\, \mu(dx) ,$$

where $x' = x - d_k$. But for all x (mod μ)

$$\varrho_\mu(d_k, x)\, \varrho_\mu(x - d_k - V_k^{-1}(x - d_k), x - d_k) = \varrho_\mu(x - V_k^{-1}(x - d_k), x) .$$

Hence,

$$\int f(d_k + V_k x)\, \mu(dx) = \int f(x)\, \varrho_\mu(x - V_k^{-1}(x - d_k), x)\, |\det V_k|\, \mu(dx) . \tag{2}$$

Let the function f be different from zero only for $x \in \Delta_k$, where $\Delta_k = \{x : x = d_k + V_k x, x \in D_k\}$. Then

$$\int f(d_k + V_k x)\, \mu(dx) = \int f(U(x))\, \mu(dx) .$$

Moreover, for $x \in \Delta_k$, $V_k^{-1}(x - d_k) = U^{-1}(x)$. Finally, $\det V_k = \det U'(U^{-1}(x))$ if x is an interior point of Δ_k. If $x \in \Delta_k \cap \Delta_j$, then $V_k^{-1}(x - d_k) = V_j^{-1}(x - d_j)$. Let $H_{k,j}$ be the subspace of those x for which $V_k^{-1} x = V_j^{-1} x$. We will assume that for $k \neq j$ either $V_k \neq V_j$ or $d_k \neq d_j$ (otherwise D_k and D_j could be combined into a single set). Then $H_{k,j}$ will be a proper subset of X. If $\mu(H_{k,j})$ were positive, then we would have $M_\mu \subset H_{k,j}$ which contradicts the fact that M_μ is dense in X.

Hence, we also have

$$\mu(\{x\colon V_k^{-1}(x-d_k) = V_j^{-1}(x-d_j)\}) = 0 .$$

This means that the set of x for which $U'(U^{-1}(x))$ is not defined has μ-measure 0, whereby

$$U'(U^{-1}(x)) = \sum_{k=1}^{m} \chi_{D_k}(x)\, V_k \qquad (\operatorname{mod} \mu) .$$

Then (2) can be written as

$$\int f(U(x))\, \mu(dx) = \int f(x)\, \varrho_\mu(x - U^{-1}(x), x)\, |\det U'(x)|^{-1}\, \mu(dx) . \quad (3)$$

Since any $\mathfrak{B}$-measurable function f can be represented in the form

$$f(x) = \sum_{1}^{m} f_k(x) ,$$

where $f_k(x)$ differs from zero only on the set Δ_k, (3) is satisfied for any function f (under the condition that $U(x)$ is a polygonal map from X into X for which $U(x) - x \in L$ for all x). We construct a sequence of polygonal operators $U_n(x)$ such that for all x (mod μ)

$$\lim_{n\to\infty} U_n(x) = U(x) ,$$

$$\lim_{n\to\infty} \det U_n'(U_n^{-1}(x)) = \det U'(U^{-1}(x))$$

and $U_n(x) - x \in L$ for all x. Choose an arbitrary orthonormalized basis $a_1, \ldots, a_N$ in L. The operator $U(x)$ is completely determined by the functions $\alpha_k(x) = (U(x) - x, a_k)$. The (numerical) function $\alpha(x)$ is called *polygonal* if one can determine closed sets D_k having no common boundary points, linear functionals $l_k(x)$ and numbers α_k such that

$$\alpha(x) = l_k(x) + \alpha_k \quad \text{for} \quad x \in D_k .$$

$U_n(x)$ is a polygonal operator iff the functions $\alpha_k^{(n)}(x) = (U_n(x) - x, a_k)$ are polygonal ($U_n(x)$ satisfies $U_n(x) - x \in L$). We will denote by $d\alpha_k(x, y)$ the differential of α_k at the point x; y is the argument of the differential. Then, if $\alpha_k(x)$ is related to $U(x)$ in the way described above, we have

$$\det U(x) = \det (d\alpha_k(x, a_j)) ; \qquad k, j = 1, \ldots, N ;$$

here, $(d\alpha_k(x, a_j))$ is an $N \times N$-matrix with elements $d\alpha_k(x, a_j)$. Hence, to construct the indicated sequence of polygonal operators $U_n(x)$ it is sufficient to construct N sequences of polygonal functions $\alpha_k^{(n)}(x)$ such that for $k = 1, \ldots, N$

$$\lim_{n\to\infty} \alpha_k^{(n)}(x) = \alpha_k(x) \quad \text{and} \quad \lim_{n\to\infty} d\alpha_k^{(n)}(x, g_j) = d\alpha_k(x, g_j) \quad \text{for } j = 1, \ldots, N$$

and all x (mod μ). To construct the functions $\alpha_k^{(n)}(x)$ we choose first a compact K for which $\mu(K) > 1 - \varepsilon_n$. Since the functions $\alpha_k(x)$ and $d\alpha_k(x, a_j)$ are uniformly continuous on the compact K, for the given ε_n we can choose a δ such that for $|x' - x''| < \delta$

$$|\alpha_k(x') - \alpha_k(x'')| < \varepsilon_n \quad \text{and} \quad |d\alpha_k(x', a_j) - d\alpha_k(x'', a_j)| < \varepsilon_n .$$

Choose a δ-net $x_1, \ldots, x_r$ in K and set

$$l_j(x) = \alpha_k(x_j) + d\alpha_k(x_j, x - x_j) .$$

Since $||d\alpha_k(x, \cdot)||$ is bounded for $x \in K$, we will assume that δ has been chosen small enough so that $\delta||d\alpha_k(x, \cdot)|| < \varepsilon_n$ for all $x \in K$. Then, for $|x - x_j| < \delta$

$$|l_j(x) - \alpha_k(x)| \leq |\alpha_k(x) - \alpha_k(x_j)| + ||d\alpha_k(x_j, \cdot)|| \cdot |x - x_j| \leq 2\,\varepsilon_n$$

and

$$|dl_j(x, a_i) - d\alpha_k(x, a_i)| = |d\alpha_k(x_j, a_i) - d\alpha_k(x, a_i)| < \varepsilon_n .$$

For $x \in \bigcup_{j=1}^{r} \{x: |x - x_j| < \delta\}$ put $\alpha_k^{(n)}(x) = l_j(x)$ if $|\alpha_k(x) - l_j(x)| \leq$ $\leq |\alpha_k(x) - l_i(x)|$ when $i \neq j$. For $x \notin \bigcup_{j=1}^{r} \{x: |x - x_j| < \delta\}$ we define $\alpha_k^{(n)}(x)$ in such a way that it is a polygonal function. Then $\alpha_k^{(n)}(x)$ will satisfy all of our requirements provided that $\varepsilon_n \to 0$ for $n \to \infty$, and we see that

$$U_n(x) = \sum_{k=1}^{N} \alpha_k^{(n)}(x) \cdot a_k$$

will be the sought-for sequence of polygonal transformations. Putting $U_n(x)$ into (3) and then letting $n \to \infty$, we find for nonnegative continuous f the inequality

$$\int f(U(x))\, \mu(dx) \geq \int f(x)\, \varrho_\mu(x - U^{-1}(x), x) \cdot |\det (U(x))' (U^{-1}(x))|\, \mu(dx) . \tag{4}$$

By similar arguments we obtain

$$\int f(U^{-1}(x))\, \mu(dx) \geq \int f(x)\, \varrho_\mu(x - U(x), x)\, |\det (U(x))' (U(x))|\, \mu(dx) . \tag{5}$$

In exactly the same way that (10) was obtained from (12) and (14) in Theorem 3, § 25 we can derive (3) from (4) and (5) for an operator $U(x)$ satisfying the conditions of the theorem provided that $U(x) - x \in L$ for all $x \in X$.

Now assume that $U(x)$ satisfies the hypotheses the theorem. Set $U_n(x) = x + P_n\,(U(x) - x)$. Then, by what we have already proved,

$$\int f(U_n(x))\, \mu(dx) = \int f(x)\, \varrho_\mu\big(P_n\,(x - U(x)), x\big)\, |\det U_n'(x)|^{-1}\, \mu(dx) . \tag{6}$$

Assuming that f is continuous and nonnegative we obtain (4) for the given $U(x)$ with the help of a limit passage. Replacing $U(x)$ in it by $U^{-1}(x)$, we get (5). But then we also have (3) for all $U(x)$ satisfying the hypotheses of the theorem. (3) is equivalent to (1). Since $\frac{d\nu}{d\mu} > 0$ $(\text{mod}\,\mu)$, $\nu \sim \mu$. □

We obtain from this theorem conditions for the equivalence of a measure ν obtained from a Gaussian measure μ by means of the transformation $U(x)$. In the course of establishing Theorem 4, § 25 we proved that for a Gaussian measure μ, Conditions 1) and 2) of Theorem 1 (of this section) are satisfied. If B is the correlation operator of the measure μ and its mean is 0, then $\varrho_\mu(a, x)$ is defined for all a and x for which the series

$$\sum_{k=1}^{\infty} \frac{(a, e_k)\cdot(x, e_k)}{\lambda_k} \quad (\text{mod}\,\mu) \quad \text{and} \quad \sum_{k=1}^{\infty} \frac{(a, e_k)^2}{\lambda_k},$$

converge, where e_k and λ_k are the eigenvectors and eigenvalues of B. Exactly as in Theorem 4, § 25, we can consider convergence in measure μ instead of convergence $(\text{mod}\,\mu)$. We have thus established

Theorem 2. *If μ is Gaussian with mean 0 and correlation operator B having eigenvectors e_k and eigenvalues λ_k, and $U(x)$ is an invertible, continuous and continuously differentiable mapping from X into X for which the following conditions hold:*

1. *for all x $(\text{mod}\,\mu)$, $\det U'(x)$ exists and $\det U'(x) \neq 0$; the transformations $U_n(x) = x + P_n\,(U(x) - x)$ and $V_n(x) = x + P_n\,(U^{-1}(x) - x)$ are invertible and*

$$\lim_{n\to\infty} |\det U_n(x)| = |\det U'(x)|\,, \quad \lim_{n\to\infty} |\det V_n'(x)| = \frac{1}{|\det U'(x)|}\,;$$

2. *for all x $(\text{mod}\,\mu)$*

$$\sum_{k=1}^{\infty} \frac{(x - U^{-1}(x), e_k)^2}{\lambda_k} < \infty \quad \textit{and} \quad \sum_{k=1}^{\infty} \frac{(x - U(x), e_k)^2}{\lambda_k} < \infty$$

and the series

$$\sum_{k=1}^{\infty} \frac{(x - U^{-1}(x), e_k)\,(x, e_k)}{\lambda_k} \quad \textit{and} \quad \sum_{k=1}^{\infty} \frac{(x - U(x), e_k)\,(x, e_k)}{\lambda_k}$$

converge in measure μ,
then, if $\nu(A) = \mu\big(U^{-1}(A)\big)$, we have $\nu \sim \mu$ and

$$\frac{d\nu}{d\mu}(x) = |\det U'(x)|^{-1} \exp\left\{\sum_{k=1}^{\infty} \frac{2\,(x - U^{-1}(x), e_k)\,(x, e_k) - (x - U^{-1}(x), e_k)^2}{\lambda_k}\right\}. \tag{7}$$

§ 27. Surface Integrals

Let a measure μ be given on $(X, \mathfrak{B})$. Our goal in this section is to associate with each sufficiently smooth surface $S \subset X$ (of co-dimension 1) a measure μ^S concentrated on S and related to μ in the same way as, for example, the Lebesgue volume and area of a surface are related in a finite-dimensional space. This connection must be made more precise. It can be understood in one of the following two ways:

I. Let S^ε denote the set of x whose distance from S does not exceed ε and let $f(x)$ be a continuous function defined for $x \in S^\varepsilon$ for some $\varepsilon > 0$. We set

$$\int f(x)\, \mu^S(dx) = \lim_{\varepsilon \to 0} \frac{1}{2\,\varepsilon} \int_{S^\varepsilon} f(x)\, \mu(dx)\,. \tag{1}$$

II. In the same notation we could also set

$$\int f(x)\, \mu^S(dx) = \lim_{\varepsilon \to 0} \frac{1}{\mu(S^\varepsilon)} \int_{S^\varepsilon} f(x)\, \mu(dx)\,. \tag{2}$$

Definition I is a direct generalization of the Lebesgue surface integral. If the integral in I is defined for all continuous bounded f and is not identically equal to zero, then the integral in II is also defined and

$$(\mathrm{II})\colon \int f(x)\, \mu^S(dx) = [(\mathrm{I})\colon \int f(x)\, \mu^S(dx)] \cdot [\int 1 \cdot \mu^S(dx)]^{-1}\,. \tag{3}$$

Definition II is thus more general. It is closely connected with the notion of a conditional measure for μ and therefore more closely reflects the connection between μ^S and μ. However, if the integral II exists for all continuous bounded functions and if

$$k = \lim_{\varepsilon \to 0} \frac{1}{2\,\varepsilon} \mu(S^\varepsilon) = (\mathrm{I})\colon \int 1 \cdot \mu^S(dx)\,,$$

also exists, then the integral I exists for all continuous bounded f and its value is given by (3).

Therefore, in the sequel we will consider the integral I. The remark following Theorem 1 § 21 implies the existence of μ^S if the limit (1) exists for all $f \in C_X$. We are interested in conditions under which μ^S exists and also want to find an expression for μ^S (along with its properties) with the help of μ.

We first consider a finite-dimensional space X and will assume that the measure μ has continuous positive density $p(x)$ w.r.t. Lebesgue measure. Regarding S, we will assume the existence at each point of a tangent hyperplane with continuously varying normal. Finally, we assume that S can be projected uniquely onto some hyperplane L.

Let $m^S(\cdot)$ be the Lebesgue surface area of S and $m(\cdot)$ Lebesgue measure in X. Then for $f \in C_X$

$$\lim_{\varepsilon\to 0} \frac{1}{2\varepsilon} \int_{S^\varepsilon} f(x)\, \mu(dx) = \lim_{\varepsilon\to 0} \frac{1}{2\varepsilon} \int_{S^\varepsilon} f(x)\, p(x)\, m(dx) = \int_S f(x)\, p(x)\, m^S(dx)\,.$$

We now express $m^S(dx)$ by means of $m^L(dx)$, Lebesgue measure on L. If $n(x)$ is the normal to S at the point x and n_L the normal to L, then setting $U_L(x) = x + \alpha\, n_L$ for $x \in L$, where α is chosen so that $U_L(x) \in S$, we obtain

$$\int_S f(x)\, p(x)\, m^S(dx) = \int_{P_L S} f(U_L(x))\, p(U_L(x)) \frac{1}{\cos\big(n(U_L(x)),\, n_L\big)} m(dx)\,. \quad (4)$$

Let μ_L denote the projection of μ onto the hyperplane. Then we also have $\mu_L \ll m^L$, and for $x \in L$

$$\frac{d\mu_L}{dm^L}(x) = \int_{-\infty}^{\infty} p(x + \tau n_L)\, d\tau\,.$$

This means that $m^L \ll \mu_L$ and

$$\frac{dm^L}{d\mu_L}(x) = \left(\int_{-\infty}^{\infty} p(x + \tau n_L)\, d\tau \right)^{-1}.$$

Inserting this in (4) and using the fact that

$$\int_{-\infty}^{\infty} \frac{p(x + \tau n_L)}{p(U_L(x))}\, d\tau = \int_{-\infty}^{\infty} \frac{p(U_L(x) - \tau n_L)}{p(U_L(x))}\, d\tau = \int_{-\infty}^{\infty} \varrho_\mu(\tau n_L,\, U_L(x))\, d\tau\,,$$

we obtain

$$\int f(x)\, \mu^S(dx) = \int_{P_L S} f(U_L(x)) \cdot \frac{1}{\int_{-\infty}^{\infty} \varrho_\mu(\tau n_L,\, U_L(x))\, d\tau \cdot \big|\big(n(U_L(x)),\, n_L\big)\big|}\, \mu_L(dx)\,. \quad (5)$$

Formula (5) is of a form that allows generalization to the infinite-dimensional case. We will find conditions under which it holds. Without loss of generality we can assume that either the surface S has no boundary or $f(x)$ vanishes in some neighborhood of the boundary of S. Assume also that the subspace L onto which S is uniquely projected is orthogonal to the vector n_L belonging to M_μ, whereby M_μ contains the line generated by the vector n_L. It follows from Theorem 1 and Remark 2 § 20 that μ is equivalent to the product measure $\mu_L \times \mu^1$, where μ_L is the projection of μ onto L, μ^1 is the projection of μ onto the subspace L_1 generated by n_L and μ^1 is absolutely continuous w.r.t. Lebesgue measure. If we denote the density of μ w.r.t. Lebesgue measure by $\sigma(x)$, than it follows from (11) § 20 that

$$\frac{d\mu_L \times \mu^1}{d\mu}(x_1 + x_2) = \sigma(x_2) \int_{-\infty}^{\infty} \varrho_\mu(\tau\, n_L, x_1 + x_2)\, d\tau, \qquad \begin{array}{l} x_1 \in L, \\ x_2 \in L_1. \end{array} \tag{6}$$

Therefore, setting

$$g(x) = \frac{d\mu}{d\mu_L \times \mu^1}(x),$$

we will have

$$\int_{S^\varepsilon} f(x)\, \mu(dx) = \int_{x_1+x_2 \in S^\varepsilon} f(x_1 + x_2) \cdot g(x_1 + x_2)\, \mu_L(dx_1)\, \mu^1(dx_2).$$

If the normal to S is uniformly continuous on S, then for small enough ε, each line parallel to n_L intersects the boundary of S^ε at no more than two points.

Define the numbers $\alpha'_\varepsilon(x) > 0$ and $\alpha''_\varepsilon(x) > 0$ by means of the relations

$$\inf_{y \in S} |y - x + \alpha'_\varepsilon(x)\, n_L| = \varepsilon, \qquad \inf_{y \in S} |y - x - \alpha''_\varepsilon(x)\, n_L| = \varepsilon.$$

Then

$$\int_{x_1 + x_2 \in S^\varepsilon} f(x_1 + x_2)\, g(x_1 + x_2)\, \mu_L(dx_1)\, \mu^1(dx_2) =$$

$$= \int_{P_L S} \int_{-\alpha'_\varepsilon(x)}^{\alpha''_\varepsilon(x)} f(U_L(x) + \tau n_L)\, g(U_L(x) + \tau n_L) \times$$

$$\times\, \sigma\Big(\big((U_L(x) - x, n_L) + \tau\big)\, n_L\Big)\, d\tau\, \mu_L(dx).$$

Taking into account the form of $g(x)$ (defined after (6)), we get

$$\int_{S_\varepsilon} f(x)\, \mu(dx) =$$

$$= \int_{P_L S} \int_{-\alpha'_\varepsilon(x)}^{\alpha''_\varepsilon(x)} f(U_L(x) + \tau n_L) \frac{1}{\int_{-\infty}^{\infty} \varrho_\mu(\alpha\, n_L, U_L(x) + \tau n_L)\, d\alpha}\, d\tau\, \mu_L(dx).$$

It is clear that when f is continuous,

$$\lim_{\varepsilon \to 0} \frac{1}{2\varepsilon} \int f(x)\, \mu(dx) =$$

$$= \lim_{\varepsilon \to 0} \int_{P_L S} f(U_L(x)) \cdot \frac{1}{2\varepsilon} \int_{-\alpha'_\varepsilon(x)}^{\alpha''_\varepsilon(x)} \frac{d\tau}{\int_{-\infty}^{\infty} \varrho_\mu(\alpha\, n_L, U_L(x) + \tau n_L)\, d\alpha}\, \mu_L(dx)$$

provided that this limit exists.

Now let the function $\left(\int_{-\infty}^{\infty} \varrho_\mu\left(\alpha n_L, U_L(x) + \tau n_L\right) d\alpha\right)^{-1}$ be continuous in τ for all x (mod μ). It is easy to see that

$$\lim_{\varepsilon\to 0} \frac{\alpha'_\varepsilon(x)}{\varepsilon} = \lim_{\varepsilon\to 0} \frac{\alpha''_\varepsilon(x)}{\varepsilon} = \frac{1}{\left|\left(n(U_L(x)), n_L\right)\right|},$$

and this limit exists uniformly if $n(x)$ is uniformly continuous on S. Hence, for any $\lambda > 1$ we have for sufficiently small $\varepsilon > 0$

$$\int\limits_{-\frac{\lambda\varepsilon}{|(n(U_L(x)), n_L)|}}^{\frac{\lambda\varepsilon}{|(n(U_L(x)), n_L)|}} \frac{d\tau}{\int_{-\infty}^{\infty} \varrho_\mu(\alpha n_L, U_L(x) + \tau n_L)\, d\alpha} \geq$$

$$\geq \int\limits_{-\alpha'_\varepsilon(x)}^{-\alpha''_\varepsilon(x)} \frac{d\tau}{\int_{-\infty}^{\infty} \varrho_\mu(\alpha n_L, U_L(x) + \tau n_L)\, d\alpha} \geq$$

$$\geq \int\limits_{-\varepsilon/\lambda|(n(U_L(x)), n_L)|}^{\varepsilon/\lambda|(n(U_L(x)), n_L)|} \frac{d\tau}{\int_{-\infty}^{\infty} \varrho_\mu(\alpha n_L, U_L(x) + \tau n_L)\, d\alpha}.$$

This implies that

$$\lim_{\varepsilon\to 0} \frac{1}{2\varepsilon} \int\limits_{S^\varepsilon} f(x)\, \mu(dx) =$$

$$= \lim_{\varepsilon\to 0} \int\limits_{P_L S} f(U_L(x)) \cdot \frac{1}{2\varepsilon} \int\limits_{-\varepsilon/|(n(U_L(x)), n_L)|}^{\varepsilon/|(n(U_L(x)), n_L)|} \frac{d\tau}{\int_{-\infty}^{\infty} \varrho_\mu(\alpha n_L, U_L(x) + \tau n_L)\, d\alpha} \cdot \mu_L(dx)$$

provided the last limit exists. Make the substitution $\tau = \tau'/|(n(U_L(x)), n_L)|$ in the inner integral on the right in this equality. Then

$$\lim_{\varepsilon\to 0} \frac{1}{2\varepsilon} \int\limits_{S^\varepsilon} f(x)\, \mu(dx) = \lim_{\varepsilon\to 0} \frac{1}{2\varepsilon} \int\limits_{-\varepsilon}^{\varepsilon} d\tau \times$$

$$\times \int\limits_{P_L S} f(U_L(x)) \frac{1}{|(n(U_L(x)), n_L)| \cdot \int\limits_{-\infty}^{\infty} \varrho_\mu\left(\alpha n_L, U_L(x) + \frac{\tau n_L}{|(n(U_L(x)), n_L)|}\right) d\alpha}\, \mu_L(dx) \tag{7}$$

(if the limit exists). Assume that the function

$$\int_{P_L S} \frac{1}{|(n(U_L(x)), n_L)| \cdot \int_{-\infty}^{\infty} \varrho_\mu\left(\alpha n_L, U_L(x) + \frac{\tau n_L}{|(n(U(x)), n_L)|}\right) d\alpha} \mu_L(dx) \qquad (8)$$

is continuous in τ at $\tau = 0$. We will show that then the limit on the left in (7) exists. Indeed, from the continuity of (8) and the continuity of the integrand in (8), it follows that for any sequence τ_n for which $\lim \tau_n = 0$, the sequence of functions

$$g_n(x) = \left[|(n(U_L(x)), n_L)| \cdot \int_{-\infty}^{\infty} \varrho_\mu\left(\alpha n_L, U_L(x) + \frac{\tau_n n_L}{|(n(U_L(x)), n_L)|}\right) d\alpha\right]^{-1}$$

is uniformly integrable w.r.t. the measure μ_L. But then the sequence of functions $f(U_L(x))\, g_n(x)$ will also be uniformly integrable since f is bounded. Hence,

$$\lim_{n\to\infty} \int_{P_L S} f(U_L(x))\, g_n(x)\, \mu_L(dx) = \int_{P_L S} \lim_{n\to\infty} f(U_L(x))\, g_n(x)\, \mu_L(dx) .$$

Thus, the function

$$\Phi(\tau) = \int_{P_L S} \frac{f(U_L(x))}{|(n(U_L(x)), n_L)| \cdot \int_{-\infty}^{\infty} \varrho_\mu\left(\alpha n_L, U_L(x) + \frac{\tau n_L}{|(n(U_L(x)), n_L)|}\right) d\alpha} \mu_L(dx)$$

is continuous in τ at $\tau = 0$ so that

$$\lim_{\varepsilon\to 0} \frac{1}{2\varepsilon} \int_{-\varepsilon}^{\varepsilon} \Phi(\tau)\, d\tau = \Phi(0) .$$

This proves the existence of the limit on the left in (7) and (5) is established. We formulate the obtained result as

Theorem 1. *Assume the surface S satisfies the following conditions:* a) *it is possible to find a subspace L onto which the surface can be uniquely projected and the normal n_L to L belongs to M_μ;* b) *at each of its interior points the surface has a tangent and the normal to the surface is uniformly continuous on it;* c) *the function of τ defined by (8), where $n(x)$ is the normal to the surface at x, $U_L(x)$ is a point on S such that $P_L U_L(x) = x$, $x \in L$, n_L is the normal to L and μ_L is the projection of μ onto L, is continuous in τ at $\tau = 0$. Then for a given μ, the surface measure μ^S*

exists and (5) *is satisfied for all* $\mathfrak{B}$*-measurable* f *defined on* S *provided the integral on the right in* (5) *exists.*

If the surface S can be decomposed into an at most countable number of components S_k for each of which we can determine a subspace L_k onto which S_k can be uniquely projected, and if the conditions of Theorem 1 hold for S_k and L_k, then the surface measure μ^S can be defined and will be σ-finite; moreover, for any nonnegative $\mathfrak{B}$-measurable function $f(x)$

$$\int f(x)\,\mu^S(dx) =$$

$$= \sum_k \int\limits_{P_{L_k} S_k} f(U_{L_k}(x)) \frac{1}{(n(U_{L_k}(x)), n_{L_k}) \cdot \int\limits_{-\infty}^{\infty} \varrho_\mu(\tau n_{L_k}, U_{L_k}(x))\,d\tau} \mu_{L_k}(dx)\,. \quad (9)$$

We have already mentioned a certain connection between measures on surfaces and conditional measures. We will illustrate this connection for surfaces of a special form. Let $g(x)$ be a continuous differentiable numerical-valued function on X. Denote by $S(\alpha)$ the surface $\{x: g(x) = \alpha\}$. Assume that the measure μ is such that for all $\alpha \in (\alpha_1, \alpha_2)$, the measure $\mu^{S(\alpha)}$ is defined on the surface $S(\alpha)$. If the derivative of $g(x)$ at x is $g'(x, y)$ (this is a linear functional of y) and $a(x)$ is a vector from X such that $g'(x, y) = (a(x), y)$ (i.e., $a(x)$ is the gradient of $g(x)$), then the tangent hyperplane to $S(\alpha)$ at x_0 is defined by the equation $(x - x_0, a(x_0)) = 0$. Suppose $a(x)$ is uniformly continuous. We can then establish that for any continuous function $f(x)$ the limit

$$\lim_{\beta\to 0} \frac{1}{\mu(V_{\alpha-\beta,\alpha+\beta})} \int\limits_{V_{\alpha-\beta,\alpha+\beta}} f(x)\,\mu(dx) = I_\alpha(f)$$

exists, where $V_{\gamma,\delta} = \{x: \gamma < g(x) < \delta\}$. This limit coincides with the value of the integral of Definition II. If we set

$$I_\alpha(f) = \int\limits_{V_{\alpha_1,\alpha}} f(x)\,\mu(dx)\,,$$

then

$$I_\alpha(f) = \frac{dI_\alpha(f)}{dI_\alpha(1)}\,, \qquad \text{i.e.,} \qquad I_\alpha(f) = \int\limits_{\alpha_1}^{\alpha} I_\beta(f)\,d_\beta\mu(V_{\alpha_1,\beta})\,.$$

From this we obtain

$$\int\limits_{V_{\alpha_1,\alpha}} f(x)\,\mu(dx) = \int\limits_{V_{\alpha_1,\alpha}} I_{g(x)}(f)\,\mu(dx)\,.$$

Hence,

$$I_{g(x)}(f) = \int f(y)\,\mu(dy, \mathfrak{B}^g/x)\,,$$

where $\mathfrak{B}^g$ is the σ-algebra generated by sets of the form $\{x: g(x) < \alpha\}$. If $I_\alpha(f)$ is continuous, then $I_\alpha(f)$ coincides for all α with an integral w.r.t. a conditional measure. Using the connection between Definitions I and II, we get the following expression:

$$\int f(y)\, \mu(dy, \mathfrak{B}^g/x) = \int\limits_{S(g(x))} f(y)\, \mu^{S(g(x))}(dy) \cdot \frac{1}{p_g(g(x))}, \tag{10}$$

where $p_g(\alpha) = \frac{d}{d\alpha} \mu\left(\{x: g(x) < \alpha\}\right)$. Formula (10) allows the calculation of the conditional measure by means of a surface integral, which can in turn be evaluated by (9).

§ 28. Gauss' Formula

In the theory of Lebesgue integration an essential role is played by formulas connecting integrals over certain regions with integrals over their boundaries. An example of such a formula is Gauss' formula: if V is a region in R^n bounded by a closed smooth surface S and $U(x)$ is a smooth vector function defined on V, then

$$\int\limits_V \operatorname{div} U(x)\, m(dx) = \int\limits_S \left(U(x), n(x)\right) m^S(dx)\,, \tag{1}$$

where $\operatorname{div} U(x) = \left(\nabla, U(x)\right) = \operatorname{tr} U'(x)$ $\left(U'(x)\right.$ is the derivative of $U(x)$ at x, and is a linear operator for each $\left.x\right)$, $n(x)$ is the outer normal to S at x, $m(dx)$ is Lebesgue measure in R^n and $m^S(dx)$ is Lebesgue measure on the surface S.

The goal of this section is the establishment of an analogous relationship between the measures μ and μ^S, where μ is some quasi-invariant measure on $(X, \mathfrak{B})$, X is a Hilbert space and μ^S is the surface measure on S constructed w.r.t. μ.

In order to clarify the possible form of Gauss' formula in this case, we first assume that X is finite-dimensional and that μ has positive differentiable density $p(x)$ w.r.t. Lebesgue measure m on $(X, \mathfrak{B})$. Since $\mu^S(dx) = p(x)\, m^S(dx)$,

$$\int\limits_S \left(U(x), n(x)\right) \mu^S(dx) = \int\limits_S \left(p(x)\, U(x), n(x)\right) m^S(dx)\,. \tag{2}$$

The function $p(x)\, U(x)$ will be differentiable if $U(x)$ is. It's easy to see that

$$[p(x)\, U(x)]'\, h = p(x)\, U'(x)\, h + \left(\operatorname{grad} p(x), h\right) U(x)\,,$$

where $h \in X$ and $\left(\operatorname{grad} p(x), a\right) = \frac{d}{dt} \left(p(x + ta)\right)\Big|_{t=0}$. Thus,

$$\operatorname{tr} [p(x)\, U(x)]' = p(x) \operatorname{tr} U'(x) + \left(\operatorname{grad} p(x), U(x)\right). \tag{3}$$

Applying (1) to the right side of (2), we get

$$\int_S (U(x)\, n(x))\, \mu^S(dx) = \int_V [p(x) + \operatorname{tr} U'(x) + (\operatorname{grad} p(x),\, U(x))]\, m(dx)\,. \tag{4}$$

Let $l_\mu(a, x)$ be the logarithmic derivative of μ in the direction a (defined in § 21). Then $\left(\frac{\operatorname{grad} p(x)}{p(x)}, a\right) = -\, l_\mu(a, x)$. Hence, the integral on the right in (4) can be written as

$$\int_V \left[\operatorname{tr} U'(x) + \left(\frac{\operatorname{grad} p(x)}{p(x)}, U(x)\right)\right] p(x)\, m(dx) =$$

$$= \int_V [p(x) \operatorname{tr} U'(x) + l_\mu(U(x), x)]\, \mu(dx)\,.$$

Finally, we obtain

$$\int_S (U(x), n(x))\, \mu^S(dx) = \int_V [\operatorname{tr} U'(x) - l_\mu(U(x), x)]\, \mu(dx)\,. \tag{5}$$

This formula also makes sense in an infinite-dimensional space X if the following conditions are satisfied:

1. S is a smooth closed surface and $n(x)$ varies continuously on S;
2. the measure μ is logarithmically differentiable w.r.t. the direction $a \in N$, where N a some linear manifold in X;
3. the measure μ and the surface S are such that μ^S is defined;
4. the function $U(x)$ acting on X into X is differentiable and $\operatorname{tr} U'(x)$ exists;
5. $U(x) \in N$ for all x (mod μ), $l_\mu(U(x), x)$ is measurable and the integral on the right in (5) exists;
6. the integral on the left in (5) exists.

The listed conditions require the existence of the integrals on the right and left in (5). Does equality hold in (5) under these conditions? We will now investigate this question.

We first assume that $U(x) = f(x) \cdot a$, where $a \in N$, $|a| = 1$ and $f(x)$ is a differentiable numerical function. We will assume that S is bounded and can be decomposed into no more than a countable number of disjoint $\mathfrak{B}$-measurable connected components in such a way that each component can be uniquely projected onto the subspace L orthogonal to a. We will assume in addition that these components of S, which we will denote by S_k, are such that the projections of the S_k onto L (denoted by D_k) either coincide or have no common points. We will say that S_k and S_j *form a pair* if $D_k = D_j$ and for all $x \in D_k$ the segment of the line $x + \tau a$ lying between S_k and S_j belongs to V. Let $V_{k,j}$ be the set of points of the form $x + \tau a$, where $x \in D_k \cap D_j$, which lie between S_k and S_j if S_k and S_j form a pair, and let $V_{k,j}$ be empty if S_k and S_j do not form a pair. Then

$$V = \bigcup_{k,j} V_{k,j} \,.$$

We will call S_k the *lower boundary* of $V_{k,j}$ if for $x \in D_k$ and $x + \tau_1 a \in S_k$, $x + \tau a \notin V_{k,j}$ for $\tau < \tau_1$. If S_k is the lower boundary of $V_{k,j}$, then S_j will be the *upper boundary* of $V_{k,j}$ (the notions upper and lower boundary are defined if $V_{k,j}$ is not empty). Consider the integral

$$\int_{V_{k,j}} [f'(x, a) - f(x)\, l_\mu(a, x)]\, \mu(dx) \,. \tag{6}$$

If a is an admissible shift for μ, then we will use the fact that $\mu \sim \mu_L \times \times \mu^1$, where μ_L is the projection of μ on L and μ^1 is its projection on the line generated by the vector a (see Theorem 1 and Remark 2, § 20). Let $\sigma(\tau)$ be the density at τa of μ^1 w.r.t. Lebesgue measure. Then, for $x \in L$

$$\frac{d\mu}{d\mu_L \times \mu^1}(x + \tau a) = \left(\sigma(\tau) \int_{-\infty}^{\infty} \varrho_\mu(\lambda a, x + \tau a)\, d\lambda\right)^{-1}$$

(see (11) § 20). Replacing the integral w.r.t. μ in (6) by an integral w.r.t. $d\mu_L \times \mu^1$, we obtain

$$\int_{x+\tau a \in V_{k,j}} [f'(x - \tau a, a) + f(x + \tau a)\, l_\mu(a, x + \tau a)] \times$$

$$\times \left(\int_{-\infty}^{\infty} \varrho_\mu(\lambda a, x + \tau a)\, d\lambda\right)^{-1} d\tau\, \mu_L(dx) =$$

$$= \int_{D_k \cap D_j} \int_{\tau_1(x)}^{\tau_2(x)} [f'(x + \tau a, a) - f(x + \tau a)\, l_\mu(a, x + \tau a)] \times$$

$$\times \left(\int_{-\infty}^{\infty} \varrho_\mu(\lambda a, x + \tau a)\, d\lambda\right)^{-1} d\tau\, \mu_L(dx) \,, \tag{7}$$

where $\tau_i(x)$ are such that $x + \tau_1(x)\, a \in S_k$, $x + \tau_2(x)\, a \in S_j$, S_k is the lower, and S_j the upper boundary of $V_{k,j}$. Starting from the formula

$$\varrho_\mu(\lambda a, x + \tau a) = \frac{\varrho_\mu((\lambda - \tau)\, a, x)}{\varrho_\mu(-\tau a, x)} \,, \tag{8}$$

which is a consequence of (2) § 19, we can write

$$\int_{-\infty}^{\infty} \varrho_\mu(\lambda a, x + \tau a)\, d\lambda = \frac{1}{\varrho_\mu(-\tau a, x)} \int_{-\infty}^{\infty} \varrho_\mu(\lambda a, x)\, d\lambda \,. \tag{9}$$

Moreover, from (20) § 21 there follows

$$\varrho_\mu(\alpha a, x) - 1 = \int_0^\alpha \varrho_\mu(\lambda a, x)\, l_\mu(a, x - \lambda a)\, d\lambda \,,$$

whence, under the assumed continuity in λ of $l_\mu(a, x - \lambda a)$ and $\varrho_\mu(\lambda a, x)$ we get

$$\frac{d}{d\alpha}\,\varrho_\mu(\alpha a, x) = \varrho_\mu(\alpha a, x)\, l_\mu(a, x - \alpha a)\,. \tag{10}$$

Using (9) and (10) for the inner integral on the right in (7), we get

$$\int\limits_{\tau_1(x)}^{\tau_2(x)} \left[f'(x + \tau a, a)\, \varrho_\mu(-\tau a, x) + f(x + \tau a)\, \frac{d}{d\tau}\, \varrho_\mu(-\tau a, x)\right] d\tau =$$

$$= \int\limits_{\tau_1(x)}^{\tau_2(x)} \frac{d}{d\tau} [f\,(x + \tau a)\, \varrho_\mu(-\tau a, x)]\, d\tau =$$

$$= f(x + \tau_2(x) \cdot a)\, \varrho_\mu(-\tau_2(x) \cdot a, x) - f(x + \tau_1(x) \cdot a)\, \varrho_\mu(-\tau_1(x) \cdot a, x)$$

(the factor $\left(\int\limits_{-\infty}^{\infty} \varrho_\mu(\lambda\, a, x)\, d\lambda\right)^{-1}$, which does not depend on τ, can be taken outside the inner integral). Hence,

$$\int\limits_{V_{k,j}} [f'(x, a) + f(x)\, l_\mu(a, x)]\, \mu(dx) =$$

$$= \int\limits_{D_k \cap D_j} \frac{f(x + \tau_2(x) \cdot a)\, \varrho_\mu(-\tau_2(x) \cdot a, x) - f(x + \tau_1(x) \cdot a)\, \varrho_\mu(-\tau_1(x) \cdot a, x)}{\int\limits_{-\infty}^{\infty} \varrho_\mu(\lambda a, x)\, d\lambda}\, \mu_L(dx)\,.$$

Using (9) again and the fact that on S_j the outer normal to S forms an acute angle with a, we have

$$\int\limits_{D_j} \frac{f(x + \tau_2(x) \cdot a)\, \varrho_\mu(-\tau_2(x) \cdot a, x)}{\int\limits_{-\infty}^{\infty} \varrho_\mu(\lambda a, x)\, d\lambda}\, \mu_L(dx) =$$

$$= \int\limits_{D_j} f(x + \tau_2(x) \cdot a)\, \big(n\,(x + \tau_2(x) \cdot a), a\big) \times$$

$$\times \frac{\mu_L(dx)}{\left|\big(n(x + \tau_2(x) \cdot a), a\big)\right| \cdot \int\limits_{-\infty}^{\infty} \varrho_\mu(\lambda a, x + \tau_2(x) \cdot a)\, d\lambda} =$$

$$= \int\limits_{S_j} f(x)\, (n(x), a)\, \mu^S(dx)\,.$$

(we used (5) § 27). We establish analogously that

$$-\int_{D_k} \frac{f(x+\tau_1(x)\cdot a)\,\varrho_\mu(-\tau_1(x)\cdot a, x)}{\int_{-\infty}^{\infty} \varrho_\mu(\lambda a, x)\,d\lambda}\,\mu_L(dx) = \int_{S_k} f(x)\,(n(x), a)\,\mu^S(dx)\,.$$

Consequently,

$$\int_{V_{k,j}} [f'(x, a) + f(x)\,l_\mu(a, x)]\,\mu(dx) = \int_{S_k \cap S_j} f(x)\,(n(x), a)\,\mu^S(dx)\,. \quad (11)$$

Summing (11) over all pairs k, j for which $V_{k,j}$ is not empty, we get

$$\int_V [f'(x, a) + f(x)\,l_\mu(a, x)]\,\mu(dx) = \int_S f(x)\,(n(x), a)\,\mu^S(dx)\,. \quad (12)$$

This will yield (5) (under certain additional assumptions) for functions $U(x)$ of the form $U(x) = f(x)\cdot a$. Since (5) depends linearly on U, it can also be established for the functions

$$U(x) = \sum_{k=1}^{n} f_k(x)\cdot a_k$$

if f_k is differentiable and $a_k \in N$ also satisfy the requirements imposed above on a. One can thus establish (5) for $P_n\,U(x)$, where P_n are operators projecting onto finite-dimensional subspaces $N_n \subset N$ such that $\bigcup_n N_n$ is dense in X. Write (5) for $P_n\,U(x)$:

$$\int_S (P_n\,U(x), n(x))\,\mu^S(dx) = \int_V [\operatorname{tr} P_n\,U'(x) + l_\mu(P_n\,U(x), x)]\,\mu(dx)\,. \quad (13)$$

It is clear that $P_n\,U(x) \to U(x)$ and $\operatorname{tr} P_n\,U'(x) \to \operatorname{tr} U'(x)$. Assume in addition that $\lim_{n\to\infty} l_\mu(P_n\,U(x), x) = l_\mu(U(x), x)$ in μ. Then, the integrands on the right and left in (13) tend to the integrands in (5). Thus, (5) can be obtained from (13) if it is possible to carry out a limit passage under the integral signs in (13).

From these considerations we get

Theorem 1. *Let $U(x)$, the measure μ and the surface S enclosing the set V satisfy Conditions 1)–6) along with the following:*

7) $M_\mu \supset N$ and the functions $\varrho_\mu(\lambda a, x)$ and $l_\mu(a, x - \lambda a)$ are continuous in λ for all x (mod μ) and $a \in N$;

8) it is possible to find a sequence of finite-dimensional subspaces $N_n \subset N$ such that $\bigcup_n N_n$ is dense in X, and if P_n is projection onto N_n, then $l_\mu(P_n\,U(x), x) \to l_\mu(U(x), x)$ in μ, the functions

$$|\operatorname{tr} P_n\,U'(x)| \qquad \text{and} \qquad |l_\mu(P_n\,U(x), x)|$$

are uniformly integrable w.r.t. μ *on the set* V *and the function* $|(P_n U(x), n(x))|$ *w.r.t.* μ^S on S;

9) *for all* $a \in N_n$, *an arbitrary line parallel to* a *intersects* S *at only a finite set of points.*

Then the formula of Gauss (5) *is valid.*

We remark that Condition (9) guarantees the possibility of decomposing S into an at most countable number of connected sets projected uniquely onto the subspace orthogonal to a for $a \in \bigcup_n N_n$.

Bibliographic Notes

Chapter 1.

§ 1. The scheme considered here for the construction of measures with the help of their values on cylinder sets can be generalized to an arbitrary linear space X on which some class of linear functionals L is defined. Assume that the σ-algebra $\mathfrak{B}$ is the minimal σ-algebra w.r.t. which all functionals $l \in L$ are measurable. Cylinder sets from $\mathfrak{B}$ are sets of the form $\{x: (l_1(x), \ldots, l_m(x)) \in A_m\}$, where $l_1, \ldots, l_m \in L$ and A_m is a Borel set from R^m. The collection of measures $\mu_{l_1,\ldots,l_m}(A_m) = \mu(\{x: (l_1(x), \ldots, l_m(x)) \in A_m\})$ are called the *finite-dimensional distributions* of the measure μ. The definition of a measure in infinite-dimensional linear spaces by means of finite-dimensional distributions was first given by Wiener [1] for a special measure in the space $C_{[0,1]}$. This measure has come to be called the *Wiener measure.* A. N. Kolmogorov [1] investigated this procedure for the construction of a measure corresponding to a random process in the space of all functions.

§ 2. Using finite-dimensional distributions one can define a weak distribution on $(X, \mathfrak{B})$ if X is a linear space and $\mathfrak{B}$ is a σ-algebra on X generated by the set of linear functionals. The question of when a weak distribution corresponds to some measure on $(X, \mathfrak{B})$ is one of the basic ones in the theory of measures in linear spaces. The main fact used in the proof of Lemma 1 is the weak compactness of a sphere. The lemma thus also holds for reflexive separable Banach spaces if $\mathfrak{B}$ is taken as the previous σ-algebra of Borel sets $\mathfrak{B}$ (L must be taken here as the conjugate X^* of X). For such spaces one can also reformulate Lemma 2 in an obvious way. Theorem 1 holds in an arbitrary complete separable metric space. Conditions for the extension of a weak distribution to a measure are contained in a paper of L. Gross [1]. A series of general results on the existence of a weak distribution for general linear spaces is presented in the lecture summary of Veršik and Sudakov [1].

§ 3. Characteristic functionals for measures in Banach spaces were introduced by Kolmogorov [2]. They have been investigated since then by numerous authors.

§ 4. The theorem was proved by Minlos [1] for denumerable Hilbert spaces and independently by Sazonov [1] in the formulation for Hilbert spaces considered here.

§ 5. Gaussian measures in Hilbert (and other) spaces have been studied in connection with Gaussian random processes. A rather complete theory of such measures is given in the book by Rozanov [2]. Gaussian measures on the l_p spaces and on certain other linear spaces were considered by Vahaniya [1]. Gaussian measures in general linear spaces have been studied by Veršik [1].

§ 6. The scheme for the construction of generalized measures in Hilbert space investigated here is taken from a paper by Daleckii [2].

Chapter 2.

§ 7. Measurable linear functionals on linear spaces with measure were considered by Veršik [2]. The facts presented here on measurable linear functionals were first published in the book by Gihman and Skorohod [2].

§ 8. The results of this section are taken mainly from Gihman and Skorohod [2].

§ 9—11. This material (here in considerably re-worked form) is presented in Gihman and Skorohod [2], Chapter 8. We remark that polynomial functionals w.r.t. Wiener measure as multiple stochastic integrals were constructed by Itô [1, 2]. Wiener [2] applied polynomials orthogonal w.r.t. Wiener measure to the study of nonlinear transformations of a space with Wiener measure. Measurable polynomial functionals in spaces of sequences were considered for product measures by Smolyanov [1].

Chapter 3.

§ 13. The general definition of a conditional distribution is due to Kolmogorov [1]. The existence of conditional measures for a wide class of measurable spaces has been proved by Rohlin [1]. The proof given here is specially adapted to the case of Hilbert space; however, with slight changes it can be extended to separable linear topological spaces.

§ 14. In this section we reformulate in analysis language known facts concerning special random processes called martingales and semi-martingales. The theory of these processes is contained in Doob's book [1], Chapter 7.

§ 15. Theorem 3, generalizing results of Theorems 1 and 2, is proved in somewhat different terminology by Grenander [1], Chapter 4. Theorem 6 is contained in Gihman and Skorohod [1]. The results of this section can be trivially generalized to measures in linear spaces.

§ 16. The absolute continuity of product measures was considered by Kakutani [1] who obtained a theorem analogous to Theorem 1. Theorem 2 was obtained by Grenander [1] in a different form.

§ 17. The absolute continuity of Gaussian measures has been treated in a large number of papers, many of them of a very specialized character. We mention only the main works: Segal [1], Gaek [1] and Feldman [1]. The papers of Gaek and Feldman essentially give necessary and sufficient conditions for the absolute continuity and singularity of Gaussian measures. Rozanov in [1] found a general formulation and simplified the proofs. Details are contained in Rozanov's book [2].

§ 18. The absolute continuity of mixed Gaussian measures was considered by Sytaya [1, 2]. Some general theorems, in particular, Theorem 1, are contained in Gihman and Skorohod [2].

Chapter 4.

§ 19. The general definition of an admissible shift and the simplest properties of admissible shifts for measures corresponding to random processes, as well as the simplest properties of sets of admissible shifts are to be found in Pitcher [1]. Part of the material of this section is taken from Skorohod [1]. The impossibility of finding measures in infinite-dimensional spaces all of whose shifts are admissible was established by Girsanov and Mityagin [1] and in a more general case by Sudakov [1]. Theorem 1 generalizes a result of Mityagin [1]. Theorem 2 can be found in Veršik [2].

§ 20. Admissible directions for measures corresponding to random processes have been considered by Pitcher [1]. Theorem 1 is a generalization of a theorem formulated by Veršik [2].

§ 21. Another definition of the derivative of a measure w.r.t. a direction can be found in Averbuh, Smolyanov and Fomin [1]. They require the existence of (1) for all bounded measurable functions. Then the derivative turns out to be everywhere absolutely continuous w.r.t. the original measure.

§ 22. Theorem 2 is a slight generalization of Theorem 7 in Skorohod [1].

§ 23. Initially, the term quasi-invariance was related to measures for which all shifts were admissible. After the lack of such measures in infinite-dimensional spaces was established, Gel'fand proposed the term for measures having a "sufficiently rich" set of admissible shifts. Here we call *quasi-invariant* measures whose set of admissible shifts contains a linear manifold dense in X. This section contains results of § 3 in Skorohod [1] in re-worked and corrected form.

Chapter 5.

§§ 24—26. Transformations of the Wiener measure were first investigated by Cameron and Martin [1, 2]. These were the first papers treating questions of absolutely continuity of measures in infinite-dimensional spaces under transformations of the spaces. They also served to give impulse to further investigation of the absolute continuity of measures in infinite-dimensional spaces. Transformed Gaussian measures in Hilbert space are treated by Baklan and Šatašvili [1]. A general formula for the density of a transformed measure w.r.t. an original measure is quoted in Gihman and Skorohod [1]; its proof is contained in [2], Chapter 7 by the same authors. The formula is proved here under more general conditions.

§§ 27, 28. Here we present with proofs results formulated in Skorohod [2]. Green's formula for Gaussian measures in terms of conditional mathematical expectations is established in Stengle [1].

Bibliography

Averbuh, V. I., Smolyanov, O. G., Fomin, S. V.

1. Generalized functions and differential equations in linear spaces. I. Differentiable measures. Trudy Mosk. Mat. Ob. **24**, 133–174 (1971) [in Russian].

Baklan, V. V., Šatašvili, A. D.

1. Conditions for the absolute continuity of probability measures corresponding to Gaussian random variables in a Hilbert Space. Dopovidi A. N. Ukr. R.S.R., **9**, 1115–1117 (1965) [in Russian].

Cameron, R. H., Martin, W. T.

1. Transformations of Wiener integrals under translations. Ann. Math. **45**, 386–396 (1944).
2. The transformation of Wiener integrals by nonlinear transformations. Trans. Am. Math. Soc. **66**, 253–283 (1949).

Daleckii, Ju. L.

1. Continual integrals related to operator evolution equations. Usp. Mat. Nauk. **17**, 3–115 (1962) [in Russian].
2. Infinite-dimensional elliptic operators and related parabolic equations. Usp. Mat. Nauk. **22**, 3–54 (1967) [in Russian].

Doob, J. L.

1. Stochastic processes. New York: Wiley 1953.

Feynman, R. P., Hibbs, A. R.

1. Quantum mechanics and path integrals. New York: McGraw-Hill 1965.

Feldman, J.

1. Equivalence and perpendicularity of Gaussian processes. Pac. J. Math. **8**, 699–708 (1958).

Gel'fand, I. M., Yaglom, A. M.

1. Integration in function spaces and its application to quantum physics . Usp. Mat. Nauk. **2**, 77–114 (1956) [in Russian].

Gihman, I. I., Skorohod, A. V.

1. On the densities of probability measures in function spaces. Usp. Mat. Nauk. **21**, 83–152 (1966) [in Russian].
2. Theory of random processes. Philadelphia: Saunders 1969.

Girsanov, I. V., Mityagin, B. S.

1. Quasi-invariant measures and linear topological spaces. Naučn. Dokl. Vys. Škol. **2**, 5–10 (1959) [in Russian].

Grenander, U.

1. Stochastic processes and statistical inference. Ark. Mat. **1**, 195–277 (1950).
2. Probabilities on algebraic structures. New York: Wiley 1963.

Gross, L.

1. Harmonic analysis in Hilbert space. Mem. Amer. Math. Soc. 46, 1—61 (1963).

Halmos, P.

1. Measure theory. New York: Van Nostrand 1950.

Hardy, G. H., Littlewood, J. E., Pólya, G.

1. Inequalities. Cambridge: Cambridge Univ. Press. 1959.

Kac, M.

1. On the distribution of certain Wiener functionals. Trans. Amer. Math. Soc. 65, 1—13 (1949).
2. On some connections between probability theory and differential and integral equations. Proc. 2nd Berkeley Sym. on Math. Stat. and Prob. pp. 189—215. Berkeley: Univ. Calif. 1951.
3. Probability and related topics in physical sciences. New York: Interscience 1959.

Kakutani, S.

1. On equivalence of infinite product measures. Ann. Math. 9, 214—224 (1948).

Kolmogorov, A. N.

1. Foundations of the theory of probability. New York: Chelsea 1956.
2. La transformation de Laplace dans les espaces linéaires. Comp. Rend. 200, 1717, (1935).

Minlos, R. A.

1. Generalized random processes. Trudy Mosk. Mat. Ob. 8, 497—518 (1959) [in Russian].

Mityagin, B. S.

1. A Remark on a quasi-invariant measure. Usp. Mat. Nauk. 16, 191—193 (1961) [in Russian].

Mourier, E.

1. Elements aléatoires dans un espace de Banach. Ann. Inst. Henri Poincaré. 13 (1953).

Pitcher, T. S.

1. The admissible mean values of a stochastic process. Trans. Amer. Math. Soc. 108, 538—546 (1963).

Prohorov, Ju. A.

1. Convergence of random processes and limit theorems in probability theory. Teor. Ver. i ee Prim. 1, 177—238 [in Russian].

Rohlin, V. A.

1. On the fundamental concepts of measure theory. Mat. Sbornik, 25, 107—150 (1949) [in Russian].

Rozanov, Ju. A.

1. On the density of one Gaussian measure w.r.t. another. Teor. Ver. i ee Prim. 7, 84—89 (1962) [in Russian].
2. Infinite-dimensional Gaussian distributions. Tr. Mat. Inst. Steklov (1966) [in Russian].

Sazonov, V. V.

1. A remark on characteristic functionals. Teor. Ver. i ee Prim. 3, 201—205 (1958) [in Russian].

Segal, I. E.

1. Distributions in Hilbert spaces and canonical systems of operators. Trans. Amer. Math. Soc. 88, 12—41 (1958).

Šilov, G. E., Fan Doc Tin.
1. The integral of a measure and the derivative on linear spaces, Moscow: Nauka 1967 [in Russian].
Skorohod, A. V.
1. On admissible shifts of measures in Hilbert Space. Teor. Ver. i ee Prim. **15**, 577—598 (1970) [in Russian].
2. Surface integrals and Green's formula in Hilbert Space; in Teor. Ver. i Mat. Statistika, **2**, pp. 172—175. Kiev: Izdvo Kiev. Univ. 1970 [in Russian].
Smolyanov, O. G.
1. On measurable multilinear and power functionals in some linear spaces with measure. Dokl. Ak. Nauk SSSR. **170**, 526—529 (1956) [in Russian].
Stengle, G.
1. A divergence theorem for Gaussian Stochastic process expectations. J. Math. Anal. Appl. **21**, 537—546 (1968).
Sudakov, V. N.
1. Linear spaces with quasi-invariant measure. Dokl. Ak. Nauk. SSSR, **127**, 524—525 (1959) [in Russian].
Sytaya, G. N.
1. On admissible shifts of weighted Gaussian measures. Teor. Ver. i ee Prim. **15**, 527—531 (1969) [in Russian].
2. On the density of weighted Gaussian measures under admissible shifts; in Teor. Ver. i Mat. Statistika. **2**, pp. 193—204. Kiev: Izdvo. Kiev. Univ. (1970) [in Russian].
Vahaniya, N. N.
1. Some problems in the theory of stochastic proccesses. Trudy Vyč. Centra Ak. Nauk. Gruz. SSSR V-**1**, 5—32 (1965). [in Russian].
Veršik, A. M.
1. The general theory of Gaussian measures in linear spaces. Usp. Mat. Nauk. **19**, 210—213 (1964) [in Russian].
2. Duality in the theory of measures in linear spaces. Dokl. Ak. Nauk. **170**, 497—500 (1966) [in Russian].
Veršik, A. M., Sudakov, V. N.
1. Probability measures in infinite-dimensional spaces. Zapiski Naučnyh Seminarov, Moscow Univ. **12**, 7—67 (1969) [in Russian].
Wiener, N.
1. Differential space. J. Math. Phys. **2**, 131—174 (1923).
2. Nonlinear Problems in random theory. New York: Wiley 1958.

Index

721/27/73 V/12/6

Ergebnisse der Mathematik und ihrer Grenzgebiete

1. Bachmann: Transfinite Zahlen.
2. Miranda: Partial Differential Equations of Elliptic Type.
4. Samuel: Méthodes d'Algèbre Abstraite en Géométrie Algébrique.
5. Dieudonné: La Géométrie des Groupes Classiques.
7. Ostmann: Additive Zahlentheorie. 1. Teil: Allgemeine Untersuchungen.
8. Wittich: Neuere Untersuchungen über eindeutige analytische Funktionen.
11. Ostmann: Additive Zahlentheorie. 2. Teil: Spezielle Zahlenmengen.
13. Segre: Some Properties of Differentiable Varieties and Transformations.
14. Coxeter/Moser: Generators and Relations for Discrete Groups.
15. Zeller/Beekmann: Theorie der Limitierungsverfahren.
16. Cesari: Asymptotic Behavior and Stability Problems in Ordinary Differential Equations.
17. Severi: Il teorema di Riemann-Roch per curve, superficie e variatà questioni collegate.
18. Jenkins: Univalent Functions and Conformal Mapping.
19. Boas/Buck: Polynomial Expansions of Analytic Functions.
20. Bruck: A Survey of Binary Systems.
21. Day: Normed Linear Spaces.
23. Bergmann: Integral Operators in the Theory of Linear Partial Differential Equations.
25. Sikorski: Boolean Algebras.
26. Künzi: Quasikonforme Abbildungen.
27. Schatten: Norm Ideals of Copletely Continuous Operators.
30. Beckenbach/Bellmann: Inequalities.
31. Wolfowitz: Coding Theorems of Information Theory.
32. Constantinescu/Cornea: Ideale Ränder Riemannscher Flächen.
33. Conner/Floyd: Differentiable Periodic Maps.
34. Mumford: Geometric Invariant Theory.
35. Gabriel/Zisman: Calculus of Fractions and Homotopy Theory.
36. Putnam: Commutation Properties of Hilbert Space Operators and Related Topics.
37. Neumann: Varieties of Groups.
38. Boas: Integrability Theorems for Trigonometric Transforms.
39. Sz.-Nagy: Spektraldarstellung linearer Transformationen des Hilbertschen Raumes.
40. Seligman: Modular Lie Algebras.
41. Deuring: Algebren.
42. Schütte: Vollständige Systeme modaler und intuitionistischer Logik.
43. Smullyan: First-Order Logic.
44. Dembowski: Finite Geometries.
45. Linnik: Ergodic Properties of Algebraic Fields.

46. Krull: Idealtheorie.
47. Nachbin: Topology on Spaces of Holomorphic Mappings.
48. A. Jonescu Tulcea/C. Ionescu Tulcea: Topics in the Theory of Lifting.
49. Hayes/Pauc: Derivation and Martingales.
50. Kahane: Séries de Fourier absolument convergentes.
51. Behnke/Thullen: Theorie der Funktionen mehrerer komplexer Veränderlichen.
52. Wilf: Finite Sections of Some Classical Inequalities.
53. Ramis: Sous-ensembles analytiques d'une variété banachique complexe.
54. Busemann: Recent Synthetic Differential Geometry.
55. Walter: Differential and Integral Inequalities.
56. Monna: Analyse non-archimedienne.
57. Alfsen: Compact Convex Sets and Boundary Integrals.
58. Greco/Salmon: Topics in m-Adic Topologies.
59. López de Medrano: Involutions on Manifolds.
60. Sakai: C*-Algebras and W*-Algebras.
61. Zariski: Algebraic Surfaces.
62. Robinson: Finiteness Conditions and Generalized Soluble Groups, Part 1.
63. Robinson: Finiteness Conditions and Generalized Soluble Groups. Part 2.
64. Hakim: Topos annelés et chémas relatifs.
65. Browder: Surgery on Simply-Connected Manifolds.
66. Pietsch: Nuclear Locally Convex Spaces.
67. Dellacherie: Capacités et processus stochastiques.
68. Raghunathan: Discrete Subgroups of Lie Groups.
69. Rourke/Sanderson: Introduction to Piecewise-Linear Topology.
70. Kobayashi: Transformation Groups in Differential Geometry.
71. Tougeron: Idéaux de fonctions différentiables.
72. Gihman/Skorohod: Stochastic Differential Equations.
73. Milnor/Husemoller: Symmetric Bilinear Forms.
74. Fossum: The Divisor Class Group of a Krull Domain.
75. Springer: Jordan Algebras and Algebraic Groups.
76. Wehrfritz: Infinite Linear Groups.
77. Radjavi/Rosenthal: Invariant Subspaces.
78. Bognár: Indefinite Inner Product Spaces.
79. Skorohod: Integration in Hilbert Space.
80. Bonsall/Duncan: Complete Normed Algebras.
81. Crossley/Nerode: Combinatorial Functors.